STUDENT EDITION

Grade 2

VOLUME 2

Mission 5: Add and Subtract Big Numbers
Mission 6: Equal Groups
Mission 7: Length, Money, and Data
Mission 8: Shapes, Time, and Fractions

NAME _____

© 2023 Zearn

Portions of this work, Zearn Math, are derivative of Eureka Math and licensed by Great Minds. © 2019 Great Minds. All rights reserved.

Zearn® is a registered trademark.

Printed in the U.S.A.

ISBN: 979-8-88868-895-3

Table of Contents

Mission 5

Lesson 1 3

Lesson 2 7

Lesson 3 9

Lesson 4 11

Lesson 5 13

Lesson 6 15

Lesson 7 17

Lesson 8 19

Lesson 9 21

Lesson 10 23

Lesson 11 25

Lesson 12 27

Lesson 13 29

Lesson 14 31

Lesson 15 39

Lesson 16 41

Lesson 17 43

Lesson 18 45

Lesson 19 47

Lesson 20 49

Mission 6

Lesson 1 53

Lesson 2 65

Lesson 3 67

Lesson 4 69

Lesson 5 71

Lesson 6 73

Lesson 7 75

Lesson 8 77

Lesson 9 81

Lesson 10 83

Lesson 11 87

Lesson 12 89

Lesson 13 101

Lesson 14 103

Lesson 15 107

Lesson 16 111

Lesson 17 115

Lesson 18 117

Lesson 19 119

Lesson 20 121

Mission 7

Lesson 1 125

Lesson 2 133

Lesson 3 137

Lesson 4 141

Lesson 5 145

Lesson 6 153

Lesson 7 157

Lesson 8 159

Lesson 9 161

Lesson 10 163

Lesson 11 165

Lesson 12 167

Lesson 13 169

Lesson 14 171

Lesson 15 175

Lesson 16 177

Lesson 17 185

Lesson 18 187

Lesson 19 189

Lesson 20 191

Lesson 21 195

Lesson 22 197

Lesson 23 201

Lesson 24 205

Lesson 25 209

Lesson 26 211

Mission 8

Lesson 1 219

Lesson 2 221

Lesson 3 223

Lesson 4 231

Lesson 5 233

Lesson 6 235

Lesson 7 237

Lesson 8 239

Lesson 9 243

Lesson 10 245

Lesson 11 247

Lesson 12 249

Lesson 13 253

Lesson 14 255

Lesson 15 257

Lesson 16 261

Grade 2

Mission 5

Add and Subtract Big Numbers

Lesson 1

Word Problem

The shelter rescued 27 kittens in June. In July, 11 kittens were rescued. In August, 40 more were rescued.

a. How many kittens did the shelter rescue during those 3 months?

b. If 64 of those kittens found homes by the end of August, how many still needed homes?

Name: _____   Date: _____

GRADE 2 / MISSION 5 / LESSON 1
Exit Ticket

1. Solve using the arrow way.

   a. 440 + 220 = _____

   b. 670 + _____ = 890

   c. _____ + 765 = 945

HUNDREDS PLACE VALUE CHART (FLUENCY TEMPLATE)

ones	
tens	
hundreds	

**UNLABELED PLACE VALUE HUNDREDS CHART
(CONCEPT EXPLORATION TEMPLATE)**

Lesson 2

Word Problem

Max has 42 marbles in his marble bag after he added 20 marbles at noon. How many marbles did he have before noon?

Name: _____   Date: _____

GRADE 2 / MISSION 5 / LESSON 2
Exit Ticket

Solve using place value strategies. Use the arrow way or mental math, and record your answers. You may use scrap paper if you like.

1. 760 − 500 = _____

   880 − 600 = _____

   990 − _____ = 590

2. 534 − 334 = _____

   _____ − 500 = 356

   736 − _____ = 136

Lesson 3

Word Problem

A children's library sold 27 donated books. Now, they have 48. How many books were there to begin with?

Name: _____   Date: _____

GRADE 2 / MISSION 5 / LESSON 3
Exit Ticket

1. Solve each set of problems using the arrow way.

    a. 440 + 300

       360 + 440

       440 + 380

    b. 670 + 230

       680 + 240

       250 + 660

Lesson 4

Word Problem

Diane needs 65 craft sticks to make a gift box. She only has 48. How many more craft sticks does she need?

Name: _____ Date: _____

GRADE 2 / MISSION 5 / LESSON 4
Exit Ticket

1. Solve using a simplifying strategy. Show your work if needed.

    830 − 530 = _____

    830 − 750 = _____

    830 − 780 = _____

2. Solve.

    a. 67 tens − 30 tens = _____ tens.   The value is _____.

    b. 67 tens − 37 tens = _____ tens.   The value is _____.

    c. 67 tens − 39 tens = _____ tens.   The value is _____.

Lesson 5

Word Problem

Jenny had 39 collectible cards in her collection. Tammy gave her 36 more. How many collectible cards does Jenny have now?

Name: _____ **Date:** _____

GRADE 2 / MISSION 5 / LESSON 5
Exit Ticket

1. Add by drawing a number bond to make a hundred. Write the simplified number sentence and solve.

    a. 390 + 210 = _____

    _____ + _____ = _____

    b. 798 + 57 = _____

    _____ + _____ = _____

2. Solve.

    53 tens + 38 tens = _____

Lesson 6

Word Problem

Maria made 60 cupcakes for the school bake sale. She sold 28 cupcakes on the first day. How many cupcakes did she have left?

Name: _____   Date: _____

GRADE 2 / MISSION 5 / LESSON 6
Exit Ticket

1. Draw and label a tape diagram to show how to simplify the problem. Write the new equation, and then subtract.

   a. 363 − 198 = _____ = _____

   b. 671 − 399 = _____ = _____

   c. 862 − 490 = _____ = _____

Lesson 7

Word Problem

Jeannie got a pedometer to count her steps. The first hour, she walked 43 steps. The next hour, she walked 48 steps.

a. How many steps did she walk in the first two hours?

b. How many more steps did she walk in the second hour than in the first?

Name: _____ Date: _____

GRADE 2 / MISSION 5 / LESSON 7
Exit Ticket

1. Circle one of the strategies below, and use the circled strategy to solve 490 + 463.

   a. Arrow way / Number bond

   b. Solve:

   c. Explain why you chose that strategy.

Lesson 8

Word Problem

Susan has 37 pennies.

M. J. has 55 more pennies than Susan.

a. How many pennies does M. J. have?

b. How many pennies do they have altogether?

ZEARN MATH Student Edition   G2M5 | Lesson 8

Name: _____   Date: _____

GRADE 2 / MISSION 5 / LESSON 8
Exit Ticket

1. Solve the following problems using your place value chart, place value disks, and vertical form. Bundle a ten or hundred, when necessary.

    a. 378 + 113

    b. 178 + 141

Lesson 9

Word Problem

The table represents the halftime score at a basketball game.

The red team scored 19 points in the second half.

The yellow team scored 13 points in the second half.

Team	Score
red team	63 points
yellow team	71 points

a. Who won the game?

b. By how much did that team win?

Name: _____   Date: _____

GRADE 2 / MISSION 5 / LESSON 9
Exit Ticket

1. Solve the following problems using your place value chart, place value disks, and vertical form. Bundle a ten or hundred, when necessary.

   a. 375 + 197

   b. 184 + 338

Lesson 10

Word Problem

Benjie has 36 crayons. Ana has 12 fewer crayons than Benjie.

a. How many crayons does Ana have?

b. How many crayons do they have altogether?

Name: _____ Date: _____

GRADE 2 / MISSION 5 / LESSON 10
Exit Ticket

1. Solve using vertical form, and draw disks on a place value chart. Bundle as needed.

   a. 436 + 509 = _____

   b. 584 + 361 = _____

Lesson 11

Word Problem

Mr. Arnold has a box of pencils. He passes out 27 pencils and has 45 left. How many pencils did Mr. Arnold have in the beginning?

Name: _____   **Date:** _____

GRADE 2 / MISSION 5 / LESSON 11
Exit Ticket

1. Solve using vertical form, and draw disks on a place value chart. Bundle as needed.

   a. 267 + 356 = _____

   b. 623 + 279 = _____

Lesson 12

Name: _____   Date: _____

GRADE 2 / MISSION 5 / LESSON 12
Exit Ticket

1. Choose the best strategy and solve. Explain why you chose that strategy.

   a. 467 + 298

   b. 300 + 524

Lesson 13

Word Problem

A fruit seller buys a carton of 90 apples. Finding that 18 of them are rotten, he throws them away. He sells 22 of the ones that are left on Monday. Now, how many apples does he have left to sell?

Name: _____ Date: _____

GRADE 2 / MISSION 5 / LESSON 13
Exit Ticket

Solve using mental math or vertical form with place value disks. Check your work using addition.

1. 378 – 117 = _____

2. 378 – 119 = _____

3. 853 – 433 = _____

4. 853 – 548 = _____

Lesson 14

Word Problem

Brienne has 23 fewer pennies than Alonzo. Alonzo has 45 pennies.

a. How many pennies does Brienne have?

b. How many pennies do Alonzo and Brienne have altogether?

Name: _____ Date: _____

GRADE 2 / MISSION 5 / LESSON 14
Exit Ticket

1. Solve by drawing place value disks on a chart. Then, use addition to check your work.

   a. 375 − 280 | Solve vertically or mentally | Check:

   b. 741 − 448 | Solve vertically or mentally | Check:

GRADE 2 CORE FLUENCY PRACTICE SET A

1.	10 + 2 =	21.	2 + 9 =
2.	10 + 5 =	22.	4 + 8 =
3.	10 + 1 =	23.	5 + 9 =
4.	8 + 10 =	24.	6 + 6 =
5.	7 + 10 =	25.	7 + 5 =
6.	10 + 3 =	26.	5 + 8 =
7.	12 + 2 =	27.	8 + 3 =
8.	14 + 3 =	28.	6 + 8 =
9.	15 + 4 =	29.	4 + 6 =
10.	17 + 2 =	30.	7 + 6 =
11.	13 + 5 =	31.	7 + 4 =
12.	14 + 4 =	32.	7 + 9 =
13.	16 + 3 =	33.	7 + 7 =
14.	11 + 7 =	34.	8 + 6 =
15.	9 + 2 =	35.	6 + 9 =
16.	9 + 9 =	36.	8 + 5 =
17.	6 + 9 =	37.	4 + 7 =
18.	8 + 9 =	38.	3 + 9 =
19.	7 + 8 =	39.	8 + 6 =
20.	8 + 8 =	40.	9 + 4 =

GRADE 2 CORE FLUENCY PRACTICE SET B

1.	10 + 7 =	21.	5 + 8 =
2.	9 + 10 =	22.	6 + 7 =
3.	2 + 10 =	23.	____ + 4 = 12
4.	10 + 5 =	24.	____ + 7 = 13
5.	11 + 3 =	25.	6 + ____ = 14
6.	12 + 4 =	26.	7 + ____ = 14
7.	16 + 3 =	27.	____ = 9 + 8
8.	15 + ____ = 19	28.	____ = 7 + 5
9.	18 + ____ = 20	29.	____ = 4 + 8
10.	13 + 5 =	30.	3 + 9 =
11.	____ = 4 + 13	31.	6 + 7 =
12.	____ = 6 + 12	32.	8 + ____ = 13
13.	____ = 14 + 6	33.	____ = 7 + 9
14.	9 + 3 =	34.	6 + 6 =
15.	7 + 9 =	35.	____ = 7 + 5
16.	____ + 4 = 11	36.	____ = 4 + 8
17.	____ + 6 = 13	37.	15 = 7 + ____
18.	____ + 5 = 12	38.	18 = ____ + 9
19.	8 + 8 =	39.	16 = ____ + 7
20.	6 + 9 =	40.	19 = 9 + ____

GRADE 2 CORE FLUENCY PRACTICE SET C

1.	15 − 5 =	21.	15 − 7 =
2.	16 − 6 =	22.	18 − 9 =
3.	17 − 10 =	23.	16 − 8 =
4.	12 − 10 =	24.	15 − 6 =
5.	13 − 3 =	25.	17 − 8 =
6.	11 − 10 =	26.	14 − 6 =
7.	19 − 9 =	27.	16 − 9 =
8.	20 − 10 =	28.	13 − 8 =
9.	14 − 4 =	29.	12 − 5 =
10.	18 − 11 =	30.	11 − 2 =
11.	11 − 2 =	31.	11 − 3 =
12.	12 − 3 =	32.	13 − 8 =
13.	14 − 2 =	33.	16 − 7 =
14.	13 − 4 =	34.	12 − 7 =
15.	11 − 3 =	35.	16 − 3 =
16.	12 − 4 =	36.	19 − 14 =
17.	13 − 2 =	37.	17 − 4 =
18.	14 − 5 =	38.	18 − 16 =
19.	11 − 4 =	39.	15 − 11 =
20.	12 − 5 =	40.	20 − 16 =

GRADE 2 CORE FLUENCY PRACTICE SET D

1.	12 – 2 =	21.	13 – 6 =
2.	15 – 10 =	22.	15 – 9 =
3.	17 – 11 =	23.	18 – 7 =
4.	12 – 10 =	24.	14 – 8 =
5.	18 – 12 =	25.	17 – 9 =
6.	16 – 13 =	26.	12 – 9 =
7.	19 – 9 =	27.	13 – 8 =
8.	20 – 10 =	28.	15 – 7 =
9.	14 – 12 =	29.	16 – 8 =
10.	13 – 3 =	30.	14 – 7 =
11.	_____ = 11 – 2	31.	13 – 9 =
12.	_____ = 13 – 2	32.	17 – 8 =
13.	_____ = 12 – 3	33.	16 – 7 =
14.	_____ = 11 – 4	34.	_____ = 13 – 5
15.	_____ = 13 – 4	35.	_____ = 15 – 8
16.	_____ = 14 – 4	36.	_____ = 18 – 9
17.	_____ = 11 – 3	37.	_____ = 20 – 6
18.	15 – 6 =	38.	_____ = 20 – 18
19.	16 – 8 =	39.	_____ = 20 – 3
20.	12 – 5 =	40.	_____ = 20 – 11

GRADE 2 CORE FLUENCY PRACTICE SET E

1.	12 + 2 =	21.	13 − 7 =
2.	14 + 5 =	22.	11 − 8 =
3.	18 + 2 =	23.	16 − 8 =
4.	11 + 7 =	24.	12 + 6 =
5.	9 + 6 =	25.	13 + 2 =
6.	7 + 8 =	26.	9 + 11 =
7.	4 + 7 =	27.	6 + 8 =
8.	13 − 6 =	28.	7 + 9 =
9.	12 − 8 =	29.	5 + 7 =
10.	17 − 9 =	30.	13 − 7 =
11.	14 − 6 =	31.	15 − 8 =
12.	16 − 7 =	32.	11 − 9 =
13.	8 + 8 =	33.	12 − 3 =
14.	7 + 6 =	34.	14 − 5 =
15.	4 + 9 =	35.	20 − 12 =
16.	5 + 7 =	36.	8 + 5 =
17.	6 + 5 =	37.	7 + 4 =
18.	13 − 8 =	38.	7 + 8 =
19.	16 − 9 =	39.	4 + 9 =
20.	14 − 8 =	40.	9 + 11 =

Lesson 15

Word Problem

Catriona earned 16 more stickers than Peter. She earned 35 stickers. How many stickers did Peter earn?

MaryJo earned 47 stickers. How many more does Peter need to have the same amount as MaryJo?

Name: _____ **Date:** _____

GRADE 2 / MISSION 5 / LESSON 15
Exit Ticket

1. Solve by drawing place value disks on a chart. Then, use addition to check your work.

   a. 583 − 327

hundreds	tens	ones

   Solve vertically or mentally

   Check:

   b. 721 − 485

hundreds	tens	ones

   Solve vertically or mentally

   Check:

Lesson 16

Word Problem

Will read 15 more pages than Marcy. Marcy read 38 pages. The book is 82 pages long.

a. How many pages did Will read?

b. How many more pages does Will need to read to finish the book?

ZEARN MATH Student Edition　　　　　　　　　　　　　　G2M5 | Lesson 16

Name: _____　　Date: _____

GRADE 2 / MISSION 5 / LESSON 16
Exit Ticket

1. Solve vertically or using mental math. Draw disks on the place value chart and unbundle, if needed.

   a. 604 − 143 = _____

hundreds	tens	ones

   b. 700 − 568 = _____

hundreds	tens	ones

Lesson 17

Word Problem

Colleen put 27 fewer beads on her necklace than Jenny did. Colleen put on 46 beads. How many beads did Jenny put on her necklace?

If 16 beads fell off of Jenny's necklace, how many beads are still on it?

Name: _____   Date: _____

GRADE 2 / MISSION 5 / LESSON 17
Exit Ticket

1. Solve vertically or using mental math. Draw disks on the place value chart and unbundle, if needed.

   a. 600 − 432 = _____

hundreds	tens	ones

   b. 303 − 254 = _____

hundreds	tens	ones

Lesson 18

Word Problem

Joseph collected 49 golf balls from the course. He still had 38 fewer than his friend Ethan.

a. How many golf balls did Ethan have?

b. If Ethan gave Joseph 24 golf balls, who had more golf balls? How many more?

Name: _____ Date: _____

GRADE 2 / MISSION 5 / LESSON 18
Exit Ticket

1. Choose the best strategy and solve. Explain why you chose that strategy.

   a. 400 − 265

   b. 507 − 198

Lesson 19

Name: _____   Date: _____

GRADE 2 / MISSION 5 / LESSON 19
Exit Ticket

1. Solve and explain why you chose that strategy.

   a. 400 + 590 = _____

   b. 775 − 497 = _____

Lesson 20

Name: _____ Date: _____

GRADE 2 / MISSION 5 / LESSON 20
Exit Ticket

Solve each problem using two different strategies.

1. 299 + 156 = _____

FIRST STRATEGY	SECOND STRATEGY
a.	b.

2. 547 + _____ = 841

FIRST STRATEGY	SECOND STRATEGY
a.	b.

Grade 2

Mission 6

Equal Groups

Lesson 1

Word Problem

Julisa has 12 stuffed animals. She wants to put the same number of animals in each of her 3 baskets.

    a. Draw a picture to show how she can put the animals into 3 equal groups.

    b. Complete the sentence.

Julisa put _____ animals in each basket.

GRADE 2 / MISSION 6 / LESSON 1
Exit Ticket

1. Circle groups of 4 hats.

2. Redraw the smiley faces into 2 equal groups.

2 groups of _____ = _____.

CORE FLUENCY PRACTICE SET A

Name: _____   Date: _____

1.	10 + 3 = _____	21.	7 + 9 = _____
2.	10 + 6 = _____	22.	4 + 8 = _____
3.	10 + 4 = _____	23.	5 + 9 = _____
4.	5 + 10 = _____	24.	8 + 6 = _____
5.	8 + 10 = _____	25.	7 + 5 = _____
6.	10 + 9 = _____	26.	5 + 8 = _____
7.	12 + 2 = _____	27.	8 + 3 = _____
8.	13 + 4 = _____	28.	9 + 8 = _____
9.	16 + 3 = _____	29.	6 + 5 = _____
10.	2 + 17 = _____	30.	7 + 6 = _____
11.	5 + 14 = _____	31.	4 + 6 = _____
12.	7 + 12 = _____	32.	8 + 7 = _____
13.	16 + 3 = _____	33.	7 + 7 = _____
14.	11 + 5 = _____	34.	8 + 6 = _____
15.	9 + 2 = _____	35.	6 + 9 = _____
16.	5 + 9 = _____	36.	8 + 5 = _____
17.	7 + 9 = _____	37.	4 + 7 = _____
18.	9 + 4 = _____	38.	3 + 9 = _____
19.	7 + 8 = _____	39.	6 + 6 = _____
20.	8 + 8 = _____	40.	4 + 9 = _____

CORE FLUENCY PRACTICE SET B

Name: _____   Date: _____

1.	10 + 4 = _____	21.	4 + 8 = _____
2.	10 + 9 = _____	22.	7 + 6 = _____
3.	5 + 10 = _____	23.	_____ + 4 = 11
4.	2 + 10 = _____	24.	_____ + 8 = 13
5.	11 + 4 = _____	25.	6 + _____ = 14
6.	12 + 5 = _____	26.	8 + _____ = 15
7.	16 + 2 = _____	27.	_____ = 9 + 8
8.	13 + _____ = 18	28.	_____ = 4 + 7
9.	11 + _____ = 20	29.	_____ = 7 + 8
10.	14 + 3 = _____	30.	3 + 9 = _____
11.	_____ = 3 + 16	31.	6 + 7 = _____
12.	_____ = 7 + 12	32.	8 + _____ = 13
13.	_____ = 15 + 4	33.	_____ = 7 + 9
14.	9 + 2 = _____	34.	6 + 5 = _____
15.	6 + 9 = _____	35.	_____ = 5 + 7
16.	_____ + 4 = 11	36.	_____ = 8 + 4
17.	_____ + 6 = 13	37.	15 = 8 + _____
18.	_____ + 5 = 12	38.	17 = _____ + 9
19.	8 + 8 = _____	39.	14 = _____ + 7
20.	6 + 6 = _____	40.	19 = 8 + _____

CORE FLUENCY PRACTICE SET C

Name: _____   Date: _____

1.	12 − 2 = _____	21.	16 − 9 = _____
2.	18 − 8 = _____	22.	14 − 6 = _____
3.	19 − 10 = _____	23.	16 − 8 = _____
4.	14 − 10 = _____	24.	15 − 6 = _____
5.	16 − 6 = _____	25.	17 − 8 = _____
6.	11 − 10 = _____	26.	18 − 9 = _____
7.	17 − 12 = _____	27.	15 − 7 = _____
8.	20 − 10 = _____	28.	13 − 8 = _____
9.	13 − 11 = _____	29.	11 − 3 = _____
10.	18 − 13 = _____	30.	12 − 5 = _____
11.	12 − 3 = _____	31.	11 − 2 = _____
12.	11 − 2 = _____	32.	13 − 6 = _____
13.	14 − 2 = _____	33.	16 − 7 = _____
14.	13 − 4 = _____	34.	12 − 8 = _____
15.	11 − 3 = _____	35.	16 − 13 = _____
16.	13 − 2 = _____	36.	15 − 14 = _____
17.	12 − 4 = _____	37.	17 − 12 = _____
18.	14 − 5 = _____	38.	19 − 16 = _____
19.	11 − 4 = _____	39.	18 − 11 = _____
20.	12 − 5 = _____	40.	20 − 16 = _____

CORE FLUENCY PRACTICE SET D

Name: _____ Date: _____

1.	19 − 9 = _____	21.	16 − 7 = _____
2.	12 − 10 = _____	22.	17 − 8 = _____
3.	18 − 11 = _____	23.	16 − 7 = _____
4.	15 − 10 = _____	24.	14 − 8 = _____
5.	17 − 12 = _____	25.	17 − 9 = _____
6.	16 − 13 = _____	26.	12 − 9 = _____
7.	12 − 2 = _____	27.	16 − 8 = _____
8.	20 − 10 = _____	28.	15 − 7 = _____
9.	14 − 11 = _____	29.	13 − 8 = _____
10.	13 − 3 = _____	30.	14 − 7 = _____
11.	_____ = 11 − 3	31.	13 − 9 = _____
12.	_____ = 14 − 4	32.	15 − 9 = _____
13.	_____ = 13 − 4	33.	14 − 6 = _____
14.	_____ = 11 − 4	34.	_____ = 13 − 5
15.	_____ = 12 − 3	35.	_____ = 15 − 8
16.	_____ = 13 − 2	36.	_____ = 18 − 9
17.	_____ = 11 − 2	37.	_____ = 20 − 4
18.	16 − 8 = _____	38.	_____ = 20 − 17
19.	15 − 6 = _____	39.	_____ = 20 − 11
20.	12 − 5 = _____	40.	_____ = 20 − 3

CORE FLUENCY PRACTICE SET E

Name: _____   Date: _____

1.	13 + 3 = _____	21.	11 − 8 = _____
2.	12 + 8 = _____	22.	13 − 7 = _____
3.	16 + 2 = _____	23.	15 − 8 = _____
4.	11 + 7 = _____	24.	12 + 6 = _____
5.	6 + 9 = _____	25.	13 + 2 = _____
6.	7 + 8 = _____	26.	9 + 11 = _____
7.	4 + 7 = _____	27.	6 + 8 = _____
8.	13 − 5 = _____	28.	8 + 9 = _____
9.	16 − 6 = _____	29.	7 + 5 = _____
10.	17 − 9 = _____	30.	13 − 7 = _____
11.	14 − 6 = _____	31.	15 − 8 = _____
12.	18 − 7 = _____	32.	11 − 9 = _____
13.	8 + 8 = _____	33.	12 − 3 = _____
14.	7 + 6 = _____	34.	14 − 5 = _____
15.	4 + 9 = _____	35.	13 + 6 = _____
16.	5 + 7 = _____	36.	8 + 5 = _____
17.	6 + 5 = _____	37.	4 + 7 = _____
18.	13 − 8 = _____	38.	7 + 8 = _____
19.	16 − 9 = _____	39.	4 + 9 = _____
20.	14 − 8 = _____	40.	20 − 12 = _____

Lesson 2

Word Problem

Mayra sorts her socks by color. She has 4 purple socks, 4 yellow socks, 4 pink socks, and 4 orange socks.

a. Draw groups to show how Mayra sorts her socks.

b. Write a repeated addition equation to match.

c. How many socks does Mayra have in all?

Name: _____ Date: _____

GRADE 2 / MISSION 6 / LESSON 2
Exit Ticket

1. Draw 1 more equal group.

   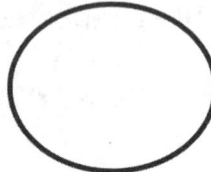

_____ + _____ + _____ + _____ = _____

4 groups of _____ = _____

2. Draw 2 groups of 3 stars. Then, write a repeated addition equation to match.

Lesson 3

Word Problem

Markers come in packs of 2. If Jessie has 6 packs of markers, how many markers does she have in all?

a. Draw groups to show Jessie's packs of markers.

b. Write a repeated addition equation to match your drawing.

c. Group addends into pairs, and add to find the total.

GRADE 2 / MISSION 6 / LESSON 3
Exit Ticket

Write a repeated addition equation to match the picture. Then, group the addends into pairs to show a more efficient way to add.

_____ + _____ + _____ + _____ = _____

_____ + _____ = _____

4 groups of _____ = 2 groups of _____

Lesson 4

Word Problem

The flowers are blooming in Maria's garden. There are 3 roses, 3 buttercups, 3 sunflowers, 3 daisies, and 3 tulips. How many flowers are there in all?

a. Draw a tape diagram to match the problem.

b. Write a repeated addition equation to solve.

GRADE 2 / MISSION 6 / LESSON 4
Exit Ticket

1. Draw a tape diagram to find the total.

   a.

   b. 3 groups of 3

   c. 2 + 2 + 2 + 2 + 2

Lesson 5

Word Problem

Mrs. White is in line at the bank. There are 4 teller windows, and 3 people are standing in line at each window.

a. Draw an array to show the people in line at the bank.

b. Write the total number of people.

Name: _____ Date: _____

GRADE 2 / MISSION 6 / LESSON 5
Exit Ticket

1. Circle groups of three. Redraw groups of three as rows and then as columns.

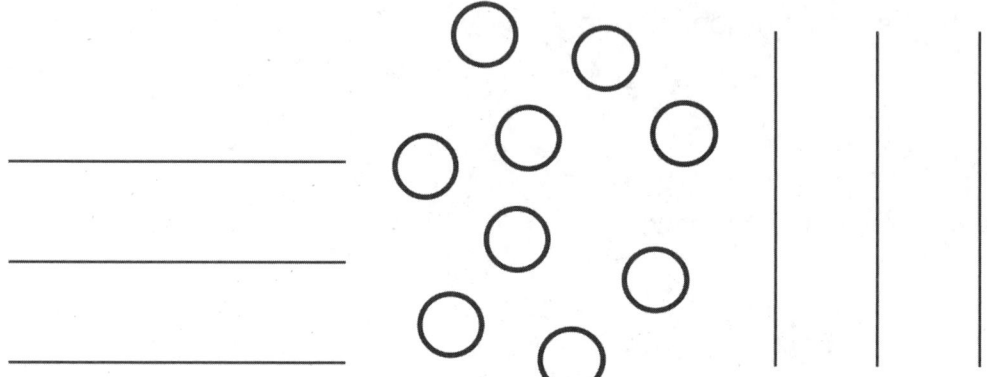

2. Complete the array by drawing more triangles. The array should have 12 triangles in all.

Lesson 6

Word Problem

Sam is organizing her greeting cards. She has 8 red cards and 8 blue cards. She puts the red cards in 2 columns and the blue ones in 2 columns to make an array.

a. Draw a picture of Sam's greeting cards in the array.

b. Write a statement about Sam's array.

Name: _____   **Date:** _____

GRADE 2 / MISSION 6 / LESSON 6
Exit Ticket

1. Use the array to answer the questions below.

   a. _____ rows of _____ = _____

   b. _____ columns of _____ = _____

   c. _____ + _____ + _____ + _____ = _____

   d. Add 1 more row. How many stars are there now? _____

   e. Add 1 more column to the new array you made in (d). How many stars are there now? _____

Lesson 7

Word Problem

Bobby puts 3 rows of tile in his kitchen to make a design. He lays 5 tiles in each row.

a. Draw a picture of Bobby's tiles.

b. Write a repeated addition equation to solve for the total number of tiles Bobby used.

Name: _____ **Date:** _____

GRADE 2 / MISSION 6 / LESSON 7
Exit Ticket

Use horizontal and vertical lines to separate the rows or columns.

1. Draw an array of X's with 3 rows of 5.

   _____ + _____ + _____ = _____

   3 rows of 5 = _____

2. Draw an array of X's with 1 more row than the above array. Write a repeated addition equation to find the total number of X's.

Lesson 8

Word Problem

Charlie has 16 blocks in his room. He wants to build equal towers with 5 blocks each.

a. Draw a picture of Charlie's towers.

b. How many towers can Charlie make?

c. How many more blocks does Charlie need to make equal towers of 5?

Name: _____ Date: _____

GRADE 2 / MISSION 6 / LESSON 8
Exit Ticket

1. Use the array of squares to answer the questions below.

    a. There are _____ squares in one row.

    b. There are _____ squares in one column.

    c. _____ + _____ + _____ = _____

    d. 3 columns of _____ = _____ rows of _____ = _____ total

2.

    a. Draw an array with 10 squares that has 5 squares in each column.

    b. Write a repeated addition equation to match the array.

Lesson 9

Name: _____ Date: _____

GRADE 2 / MISSION 6 / LESSON 9
Exit Ticket

Draw a tape diagram or an array for each word problem. Then, write a repeated addition equation to match.

1. Olivia cleans 3 cars every hour at work. She worked 4 hours on Saturday. How many cars did Olivia clean on Saturday?

2. Joshua put 5 stickers on each page in his sticker album. He filled 5 pages with stickers. How many stickers did Joshua use?

Lesson 10

Word Problem

Sandy's toy telephone has buttons arranged in 3 columns and 4 rows.

a. Draw a picture of Sandy's telephone.

b. Write a repeated addition equation to show the total number of buttons on Sandy's telephone.

Name: _____   **Date:** _____

GRADE 2 / MISSION 6 / LESSON 10
Exit Ticket

On this sheet, use square tiles to construct the following arrays with no gaps or overlaps. Write a repeated addition equation to match your construction.

1.
   a. Construct a rectangle with 2 rows of 5 tiles.

   b. Write the repeated addition equation:

   _____

2.

    a. Construct a rectangle with 5 columns of 2 tiles.

    b. Write the repeated addition equation:

    _____

Lesson 11

Word Problem

Ty bakes two pans of brownies. In the first pan, he cuts 2 rows of 8. In the second pan, he cuts 4 rows of 4.

a. Draw a picture of Ty's brownie pans.

b. Write a repeated addition equation to show the total number of brownies in each pan.

c. How many brownies did Ty bake altogether? Write an equation and a statement to show your answer.

Name: _____   Date: _____

GRADE 2 / MISSION 6 / LESSON 11
Exit Ticket

a. Construct an array with 12 square tiles.

b. Write a repeated addition equation to match the array.

_____

Lesson 12

Word Problem

Lulu made a pan of brownies. She cut them into 3 rows and 3 columns.

a. Draw a picture of Lulu's brownies in the pan.

b. Write a number sentence to show how many brownies Lulu has.

c. Write a statement about Lulu's brownies.

Extension: How should Lulu cut her brownies if she wants to equally serve 12 people? 16 people? 20 people?

ZEARN MATH Student Edition G2M6 | Lesson 12

Name: _____   Date: _____

GRADE 2 / MISSION 6 / LESSON 12
Exit Ticket

1. Draw an array of 3 columns of 3, without gaps or overlaps, starting with the square below.

CORE FLUENCY PRACTICE SET A

Name: _____   Date: _____

1.	10 + 2 = _____	21.	7 + 9 = _____
2.	10 + 7 = _____	22.	5 + 8 = _____
3.	10 + 5 = _____	23.	3 + 9 = _____
4.	4 + 10 = _____	24.	8 + 6 = _____
5.	6 + 11 = _____	25.	7 + 4 = _____
6.	12 + 2 = _____	26.	9 + 5 = _____
7.	14 + 3 = _____	27.	6 + 6 = _____
8.	13 + 5 = _____	28.	8 + 3 = _____
9.	17 + 2 = _____	29.	7 + 6 = _____
10.	12 + 6 = _____	30.	6 + 9 = _____
11.	11 + 9 = _____	31.	8 + 7 = _____
12.	2 + 16 = _____	32.	9 + 9 = _____
13.	15 + 4 = _____	33.	5 + 7 = _____
14.	5 + 9 = _____	34.	8 + 4 = _____
15.	9 + 2 = _____	35.	6 + 5 = _____
16.	4 + 9 = _____	36.	9 + 7 = _____
17.	9 + 6 = _____	37.	6 + 8 = _____
18.	8 + 9 = _____	38.	2 + 9 = _____
19.	7 + 8 = _____	39.	9 + 8 = _____
20.	8 + 8 = _____	40.	7 + 7 = _____

CORE FLUENCY PRACTICE SET B

Name: _____   Date: _____

1.	10 + 6 = _____	21.	3 + 8 = _____
2.	10 + 9 = _____	22.	9 + 4 = _____
3.	7 + 10 = _____	23.	_____ + 6 = 11
4.	3 + 10 = _____	24.	_____ + 9 = 13
5.	5 + 11 = _____	25.	8 + _____ = 14
6.	12 + 8 = _____	26.	7 + _____ = 15
7.	14 + 3 =	27.	_____ = 4 + 8
8.	13 + _____ = 19	28.	_____ = 8 + 9
9.	15 + _____ = 18	29.	_____ = 6 + 4
10.	12 + 5 = _____	30.	3 + 9 = _____
11.	_____ = 2 + 17	31.	5 + 7 = _____
12.	_____ = 3 + 13	32.	8 + _____ = 14
13.	_____ = 16 + 2	33.	_____ = 5 + 9
14.	9 + 3 = _____	34.	8 + 8 = _____
15.	6 + 9 = _____	35.	_____ = 7 + 9
16.	_____ + 5 = 14	36.	_____ = 8 + 4
17.	_____ + 7 = 13	37.	17 = 8 + _____
18.	_____ + 8 = 12	38.	19 = _____ + 9
19.	8 + 7 = _____	39.	12 = _____ + 7
20.	7 + 6 = _____	40.	15 = 8 + _____

CORE FLUENCY PRACTICE SET C

Name: _____   Date: _____

1.	13 − 3 = _____	21.	16 − 8 = _____
2.	19 − 9 = _____	22.	14 − 5 = _____
3.	15 − 10 = _____	23.	16 − 7 = _____
4.	18 − 10 = _____	24.	15 − 7 = _____
5.	12 − 2 = _____	25.	17 − 8 = _____
6.	11 − 10 = _____	26.	18 − 9 = _____
7.	17 − 13 = _____	27.	15 − 6 = _____
8.	20 − 10 = _____	28.	13 − 8 = _____
9.	14 − 11 = _____	29.	14 − 6 = _____
10.	16 − 12 = _____	30.	12 − 5 = _____
11.	11 − 3 = _____	31.	11 − 7 = _____
12.	13 − 2 = _____	32.	13 − 8 = _____
13.	14 − 2 = _____	33.	16 − 9 = _____
14.	13 − 4 = _____	34.	12 − 8 = _____
15.	12 − 3 = _____	35.	16 − 12 = _____
16.	11 − 4 = _____	36.	18 − 15 = _____
17.	12 − 5 = _____	37.	15 − 14 = _____
18.	14 − 5 = _____	38.	17 − 11 = _____
19.	11 − 2 = _____	39.	19 − 13 = _____
20.	12 − 4 = _____	40.	20 − 12 = _____

ZEARN MATH Student Edition   G2M6 | Lesson 12

CORE FLUENCY PRACTICE SET D

Name: _____   Date: _____

1.	17 – 7 = _____	21.	16 – 7 = _____
2.	14 – 10 = _____	22.	17 – 8 = _____
3.	19 – 11 = _____	23.	18 – 7 = _____
4.	16 – 10 = _____	24.	14 – 6 = _____
5.	17 – 12 = _____	25.	17 – 8 = _____
6.	15 – 13 = _____	26.	12 – 8 = _____
7.	12 – 3 = _____	27.	14 – 7 = _____
8.	20 – 11 = _____	28.	15 – 8 = _____
9.	18 – 11 = _____	29.	13 – 5 = _____
10.	13 – 5 = _____	30.	16 – 8 = _____
11.	_____ = 11 – 2	31.	14 – 9 = _____
12.	_____ = 12 – 4	32.	15 – 6 = _____
13.	_____ = 13 – 5	33.	13 – 6 = _____
14.	_____ = 12 – 3	34.	_____ = 13 – 8
15.	_____ = 11 – 4	35.	_____ = 15 – 7
16.	_____ = 13 – 2	36.	_____ = 18 – 9
17.	_____ = 11 – 3	37.	_____ = 20 – 14
18.	17 – 8 = _____	38.	_____ = 20 – 7
19.	14 – 6 = _____	39.	_____ = 20 – 11
20.	16 – 9 = _____	40.	_____ = 20 – 8

CORE FLUENCY PRACTICE SET E

Name: _____ Date: _____

1.	11 + 9 = _____	21.	13 − 7 = _____
2.	13 + 5 = _____	22.	11 − 8 = _____
3.	14 + 3 = _____	23.	15 − 6 = _____
4.	12 + 7 = _____	24.	12 + 7 = _____
5.	5 + 9 = _____	25.	14 + 3 = _____
6.	8 + 8 = _____	26.	8 + 12 = _____
7.	14 − 7 = _____	27.	5 + 7 = _____
8.	13 − 5 = _____	28.	8 + 9 = _____
9.	16 − 7 = _____	29.	7 + 5 = _____
10.	17 − 9 = _____	30.	13 − 6 = _____
11.	14 − 6 = _____	31.	14 − 8 = _____
12.	18 − 5 = _____	32.	12 − 9 = _____
13.	9 + 9 = _____	33.	11 − 3 = _____
14.	7 + 6 = _____	34.	14 − 5 = _____
15.	3 + 9 = _____	35.	13 − 8 = _____
16.	6 + 7 = _____	36.	8 + 5 = _____
17.	8 + 5 = _____	37.	4 + 7 = _____
18.	13 − 8 = _____	38.	7 + 8 = _____
19.	16 − 9 = _____	39.	4 + 9 = _____
20.	14 − 8 = _____	40.	20 − 8 = _____

Lesson 13

Word Problem

Ellie bakes a square pan of lemon bars, which she cut into nine equal pieces. Her brothers eat 1 row of her treats. Then, her mom eats 1 column.

a. Draw a picture of Ellie's lemon bars before any are eaten. Write a number sentence to show how to find the total.

b. Write an X on the bars that her brothers eat. Write a new number sentence to show how many are left.

c. Draw a line through the bars that her mom eats. Write a new number sentence to show how many are left.

d. How many bars are left? Write a statement.

Name: _____   Date: _____

GRADE 2 / MISSION 6 / LESSON 13
Exit Ticket

1. Use square tiles to complete the steps for each problem.

    Step 1: Construct a rectangle with 5 columns of 4.

    Step 2: Separate 2 columns of 4.

    Step 3: Write a number bond to show the whole and two parts.

    Step 4: Write a repeated addition sentence to match each part of the number bond.

Lesson 14

Name: _____ Date: _____

GRADE 2 / MISSION 6 / LESSON 14
Exit Ticket

1. With tiles, show 1 rectangle with 12 squares. Complete the sentences below.

   I see _____ rows of _____.

   In the exact same rectangle, I see _____ columns of _____.

PROBLEM SET

Cut out Rectangles A, B, and C. Then, cut according to directions. Answer each of the following using Rectangles A, B, and C.[1]

1. Cut out each row of Rectangle A.

    a. Rectangle A has _____ rows.

    b. Each row has _____ squares.

    c. _____ rows of _____ = _____

    d. Rectangle A has _____ squares.

2. Cut out each column of Rectangle B.

    a. Rectangle B has _____ columns.

    b. Each column has _____ squares.

    c. _____ columns of _____ = _____

    d. Rectangle B has _____ squares.

---

[1] Note: This Problem Set is used with a template of three identical 2 by 4 arrays. These arrays are labeled as Rectangles A, B, and C.

3. Cut out each square from both Rectangles A and B.

   a. Construct a new rectangle using all 16 squares.

   b. My rectangle has _____ rows of _____.

   c. My rectangle also has _____ columns of _____.

   d. Write two repeated addition number sentences to match your rectangle.

4. Construct a new array using the 24 squares from Rectangles A, B, and C.

   a. My rectangle has _____ rows of _____.

   b. My rectangle also has _____ columns of _____.

   c. Write two repeated addition number sentences to match your rectangle.

5. **Extension:** Construct another array using the squares from Rectangles A, B, and C.

   a. My rectangle has _____ rows of _____.

   b. My rectangle also has _____ columns of _____.

   c. Write two repeated addition number sentences to match your rectangle.

Lesson 15

Word Problem

Rick is filling his muffin pan with batter. He fills 2 columns of 4. One column of 4 is empty.

a. Draw to show the muffins and the empty column.

b. Write a repeated addition equation to tell how many muffins Rick makes.

GRADE 2 / MISSION 6 / LESSON 15
Exit Ticket

Shade in an array with 3 rows of 5.

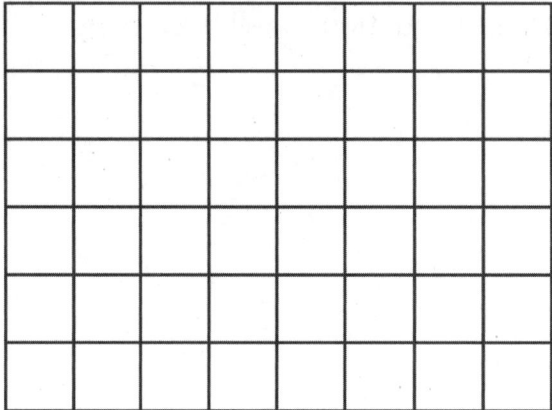

Write a repeated addition equation for the array.

_____

PROBLEM SET

1. Shade in an array with 2 rows of 3.

   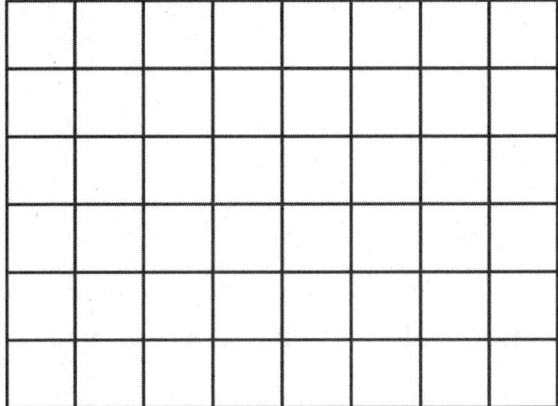

   Write a repeated addition equation for the array.

   _____

2. Shade in an array with 4 rows of 3.

   Write a repeated addition equation for the array.

   _____

3. Shade in an array with 5 columns of 4.

   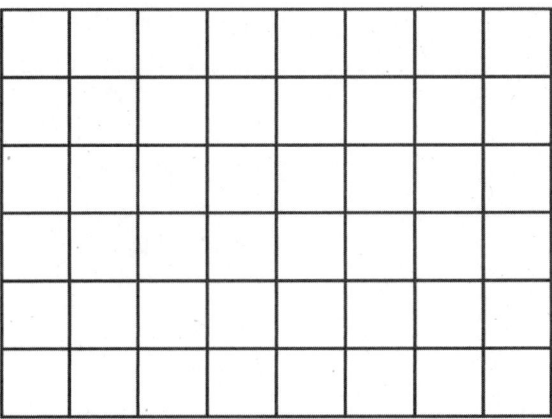

   Write a repeated addition equation for the array.

   _____

4. Draw one more column of 2 to make a new array.

Write a repeated addition equation for the array.

_____

5. Draw one more row of 4 and then one more column to make a new array.

Write a repeated addition equation for the array.

_____

6. Draw one more row and then two more columns to make a new array.

Write a repeated addition equation for the array.

_____

Lesson 16

Word Problem

Rick is baking muffins again. He filled 3 columns of 3 and left one column of 3 empty.

a. Draw a picture to show what the muffin pan looked like. Shade the columns that Rick filled.

b. Write a repeated addition equation to tell how many muffins Rick makes. Then, write a repeated addition equation to tell how many muffins would fit in the whole pan.

Name: _____ **Date:** _____

GRADE 2 / MISSION 6 / LESSON 16
Exit Ticket

Use your square tiles and grid paper to complete the following.

a. Create a design with the paper tiles you used in the lesson.

b. Shade in your design on the grid paper.

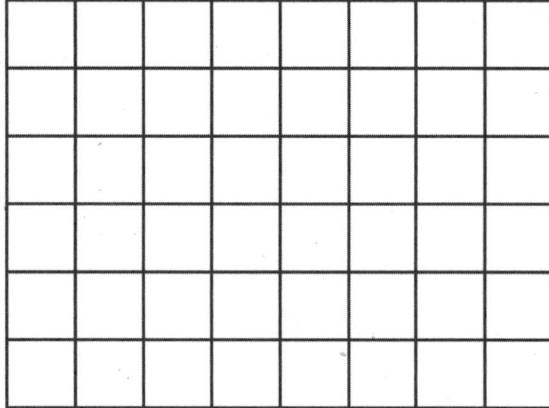

PROBLEM SET

Use your square tiles and grid paper to complete the following problems.

Problem 1

    a. Cut out 10 square tiles.

    b. Cut one of your square tiles in half diagonally.

    c. Create a design.

    d. Shade in your design on grid paper.

Problem 2

    a. Use 16 square tiles.

    b. Cut two of your square tiles in half diagonally.

    c. Create a design.

    d. Shade in your design on grid paper.

    e. Share your second design with your partner.

    f. Check each other's copy to be sure it matches the tile design.

Problem 3

    a. Create a 3 by 3 design with your partner in the corner of a new piece of grid paper.

    b. With your partner, copy that design to fill the entire paper.

Lesson 17

Word Problem

Seven students sit on one side of a lunch table. Seven more students sit across from them on the other side of the table.

  a. Draw an array to show the students.

  b. Write an addition equation that matches the array.

Three more students sit down on each side of the table.

  a. Draw an array to show how many students there are now.

  b. Write an addition equation that matches the new array.

Name: _____   Date: _____

GRADE 2 / MISSION 6 / LESSON 17
Exit Ticket

1. Draw an array for each set. Complete the sentences.

    a. 2 rows of 5

    2 rows of 5 = _____

    _____ + _____ = _____

    Circle one: 5 doubled is   even / not even.

    b. 2 rows of 3

    2 rows of 3 = _____

    _____ + _____ = _____

    Circle one: 3 doubled is   even / not even.

Lesson 18

Word Problem

Eggs come in cartons of 12. Use pictures, numbers, or words to explain whether 12 is even or not even.

Exit Ticket

GRADE 2 / MISSION 6 / LESSON 18

Redraw the following sets of dots as columns of two or 2 equal rows.

1.

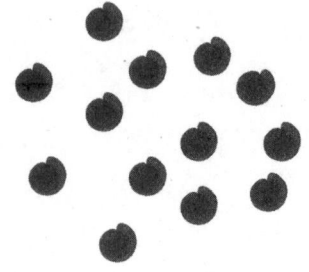

There are _____ dots.

Is _____ an even number? _____

2.

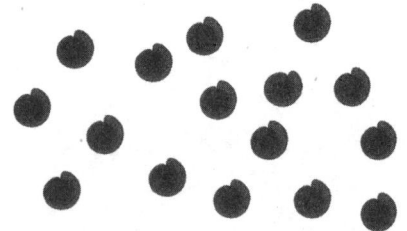

There are _____ dots.

Is _____ an even number? _____

Lesson 19

Word Problem

Eggs come in cartons of 12. Joanna's mom used 1 egg. Use pictures, numbers, or words to explain whether the amount left is even or odd.

ZEARN MATH Student Edition   G2M6 | Lesson 19

Name: _____   Date: _____

GRADE 2 / MISSION 6 / LESSON 19
Exit Ticket

1. Are the **bold** numbers even or odd? Circle the answer, and explain how you know.

   EXPLANATION

   a.
   18

   even / odd

   b.
   23

   even / odd

Lesson 20

Word Problem

Mrs. Boxer has 11 boys and 9 girls at a Grade 2 party.

a. Write the equation to show the total number of people.

b. Are the addends even or odd?

c. Mrs. Boxer wants to pair everyone up for a game. Does she have the right number of people for everyone to have a partner?

GRADE 2 / MISSION 6 / LESSON 20
Exit Ticket

Use the objects to create an array.

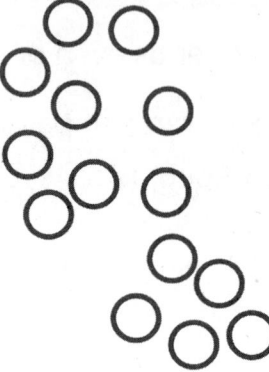

Array:

There are an   even / odd
(circle one) number of circles.

Redraw your picture with 1 *less* circle.

There are an   even / odd
(circle one) number of circles.

Grade 2

Mission 7

Length, Money, and Data

Lesson 1

Word Problem

There are 24 penguins sliding on the ice. There are 18 whales splashing in the ocean. How many more penguins than whales are there?

Name: _____   Date: _____

GRADE 2 / MISSION 7 / LESSON 1
Exit Ticket

Use the Animal Classification table to answer the following questions about the types of animals at the local zoo.

Animal Classification

Birds	Fish	Mammals	Reptiles
9	4	17	8

1. How many animals are birds, fish or reptiles? _____

2. How many more mammals are there than fish? _____

3. How many animals were classified? _____

4. How many more animals would need to be added to the chart to have 45 animals classified? _____

CORE FLUENCY PRACTICE SET A

Name _____ Date _____

1.	10 + 2 =	21.	7 + 9 =
2.	10 + 7 =	22.	5 + 8 =
3.	10 + 5 =	23.	3 + 9 =
4.	4 + 10 =	24.	8 + 6 =
5.	6 + 11 =	25.	7 + 4 =
6.	12 + 2 =	26.	9 + 5 =
7.	14 + 3 =	27.	6 + 6 =
8.	13 + 5 =	28.	8 + 3 =
9.	17 + 2 =	29.	7 + 6 =
10.	12 + 6 =	30.	6 + 9 =
11.	11 + 9 =	31.	8 + 7 =
12.	2 + 16 =	32.	9 + 9 =
13.	15 + 4 =	33.	5 + 7 =
14.	5 + 9 =	34.	8 + 4 =
15.	9 + 2 =	35.	6 + 5 =
16.	4 + 9 =	36.	9 + 7 =
17.	9 + 6 =	37.	6 + 8 =
18.	8 + 9 =	38.	2 + 9 =
19	7 + 8 =	39.	9 + 8 =
20.	8 + 8 =	40.	7 + 7 =

CORE FLUENCY PRACTICE SET B

Name _____ Date _____

1.	10 + 6 =	21.	3 + 8 =
2.	10 + 9 =	22.	9 + 4 =
3.	7 + 10 =	23.	___ + 6 = 11
4.	3 + 10 =	24.	___ + 9 = 13
5.	5 + 11 =	25.	8 + ___ = 14
6.	12 + 8 =	26.	7 + ___ = 15
7.	14 + 3 =	27.	___ = 4 + 8
8.	13 + ___ = 19	28.	___ = 8 + 9
9.	15 + ___ = 18	29.	___ = 6 + 4
10.	12 + 5 =	30.	3 + 9 =
11.	___ = 2 + 17	31.	5 + 7 =
12.	___ = 3 + 13	32.	8 + ___ = 14
13.	___ = 16 + 2	33.	___ = 5 + 9
14.	9 + 3 =	34.	8 + 8 =
15.	6 + 9 =	35.	___ = 7 + 9
16.	___ + 5 = 14	36.	___ = 8 + 4
17.	___ + 7 = 13	37.	17 = 8 + ___
18.	___ + 8 = 12	38.	19 = ___ + 9
19.	8 + 7 =	39.	12 = ___ + 7
20.	7 + 6 =	40.	15 = 8 + ___

CORE FLUENCY PRACTICE SET C

Name _____ Date _____

1.	13 − 3 =	21.	16 − 8 =
2.	19 − 9 =	22.	14 − 5 =
3.	15 − 10 =	23.	16 − 7 =
4.	18 − 10 =	24.	15 − 7 =
5.	12 − 2 =	25.	17 − 8 =
6.	11 − 10 =	26.	18 − 9 =
7.	17 − 13 =	27.	15 − 6 =
8.	20 − 10 =	28.	13 − 8 =
9.	14 − 11 =	29.	14 − 6 =
10.	16 − 12 =	30.	12 − 5 =
11.	11 − 3 =	31.	11 − 7 =
12.	13 − 2 =	32.	13 − 8 =
13.	14 − 2 =	33.	16 − 9 =
14.	13 − 4 =	34.	12 − 8 =
15.	12 − 3 =	35.	16 − 12 =
16.	11 − 4 =	36.	18 − 15 =
17.	12 − 5 =	37.	15 − 14 =
18.	14 − 5 =	38.	17 − 11 =
19	11 − 2 =	39.	19 − 13 =
20.	12 − 4 =	40.	20 − 12 =

CORE FLUENCY PRACTICE SET D

Name _____   Date _____

1.	17 − 7 =	21.	16 − 7 =
2.	14 − 10 =	22.	17 − 8 =
3.	19 − 11 =	23.	18 − 7 =
4.	16 − 10 =	24.	14 − 6 =
5.	17 − 12 =	25.	17 − 8 =
6.	15 − 13 =	26.	12 − 8 =
7.	12 − 3 =	27.	14 − 7 =
8.	20 − 11 =	28.	15 − 8 =
9.	18 − 11 =	29.	13 − 5 =
10.	13 − 5 =	30.	16 − 8 =
11.	____ = 11 − 2	31.	14 − 9 =
12.	____ = 12 − 4	32.	15 − 6 =
13.	____ = 13 − 5	33.	13 − 6 =
14.	____ = 12 − 3	34.	____ = 13 − 8
15.	____ = 11 − 4	35.	____ = 15 − 7
16.	____ = 13 − 2	36.	____ = 18 − 9
17.	____ = 11 − 3	37.	____ = 20 − 14
18.	17 − 8 =	38.	____ = 20 − 7
19	14 − 6 =	39.	____ = 20 − 11
20.	16 − 9 =	40.	____ = 20 − 8

CORE FLUENCY PRACTICE SET E

Name _____ Date _____

1.	11 + 9 =	21.	13 − 7 =
2.	13 + 5 =	22.	11 − 8 =
3.	14 + 3 =	23.	15 − 6 =
4.	12 + 7 =	24.	12 + 7 =
5.	5 + 9 =	25.	14 + 3 =
6.	8 + 8 =	26.	8 + 12 =
7.	14 − 7 =	27.	5 + 7 =
8.	13 − 5 =	28.	8 + 9 =
9.	16 − 7 =	29.	7 + 5 =
10.	17 − 9 =	30.	13 − 6 =
11.	14 − 6 =	31.	14 − 8 =
12.	18 − 5 =	32.	12 − 9 =
13.	9 + 9 =	33.	11 − 3 =
14.	7 + 6 =	34.	14 − 5 =
15.	3 + 9 =	35.	13 − 8 =
16.	6 + 7 =	36.	8 + 5 =
17.	8 + 5 =	37.	4 + 7 =
18.	13 − 8 =	38.	7 + 8 =
19.	16 − 9 =	39.	4 + 9 =
20.	14 − 8 =	40.	20 − 8 =

Lesson 2

Word Problem

Gemma is counting animals in the park. She counts 16 robins, 19 ducks, and 17 squirrels. How many more robins and ducks did Gemma count than squirrels?

GRADE 2 / MISSION 7 / LESSON 2
Exit Ticket

1. Use grid paper to create a picture graph below using data provided in the table. Then, answer the questions.

Fairview Park Zoo Animal Classification			
Birds	Fish	Mammals	Reptiles
8	4	12	5

   Title: _____

   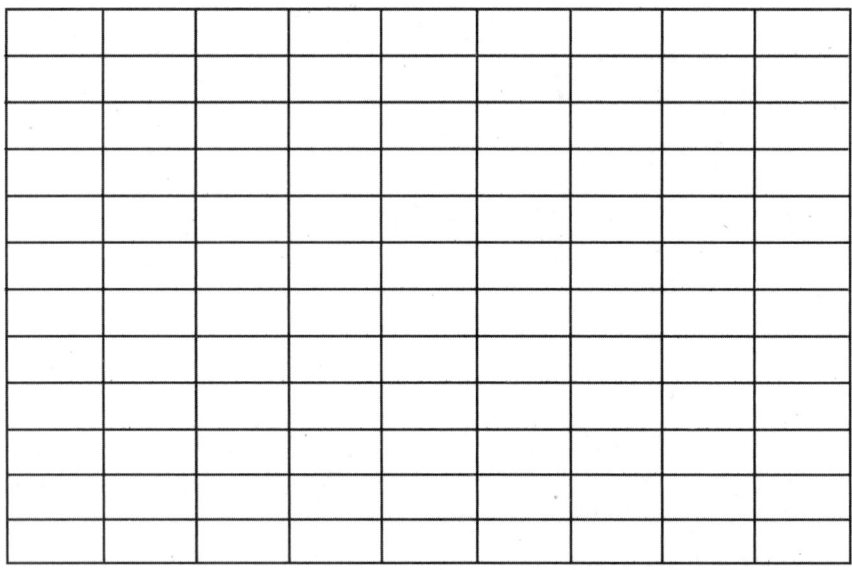

   Legend: _____

   a. How many more animals are mammals than birds? _____

   b. How many more animals are mammals and reptiles than birds and fish?
   _____

   c. How many fewer animals are fish than birds? _____

ZEARN MATH Student Edition   G2M7 | Lesson 2

VERTICAL AND HORIZONTAL PICTURE GRAPHS
(CONCEPT EXPLORATION TEMPLATE 1)

Legend: _____

Legend: _____

VERTICAL PICTURE GRAPH (CONCEPT EXPLORATION TEMPLATE 2)

_____

Legend: _____

Lesson 3

Word Problem

Number of Books Read

Jose	Laura	Linda

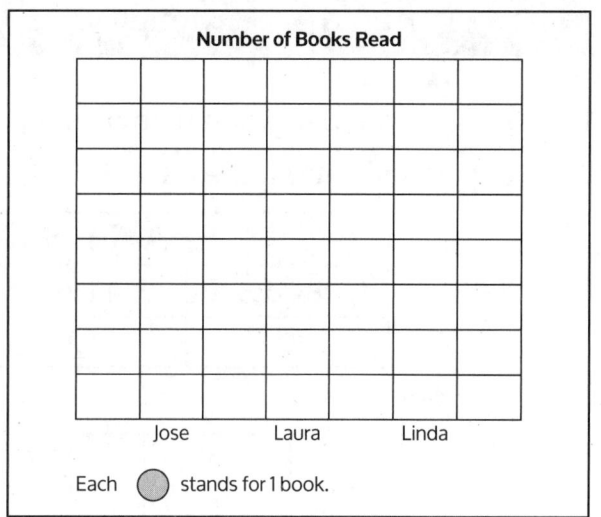

a. Use the tally chart to fill in the picture graph.

b. Draw a tape diagram to show how many more books Jose read than Laura.

c. If Jose, Laura, and Linda read 21 books altogether, how many books did Linda read?

d. Complete the tally chart and the graph.

Name: _____   Date: _____

GRADE 2 / MISSION 7 / LESSON 3
Exit Ticket

1. Complete the bar graph below using data provided in the table. Then, answer the questions about the data.

Animal Classification			
Birds	Fish	Mammals	Reptiles
7	12	8	6

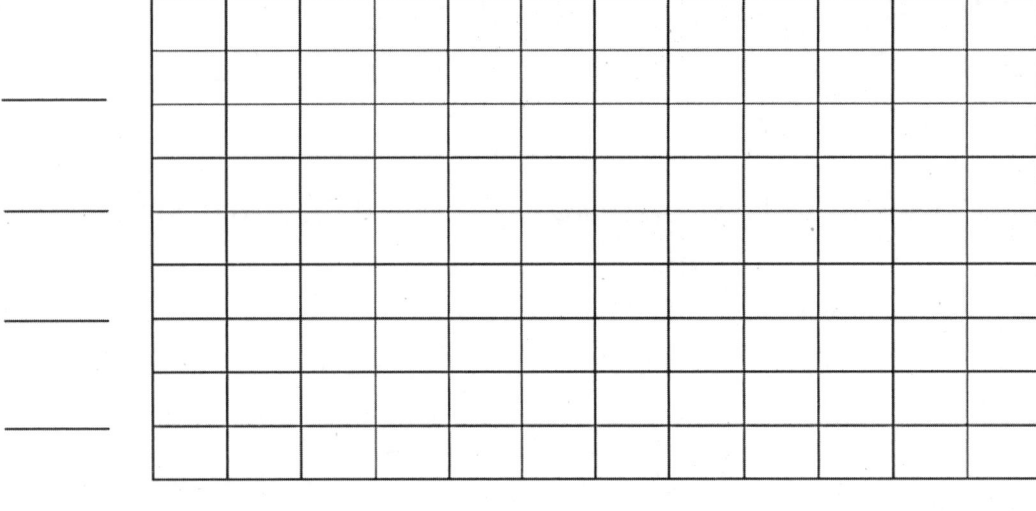

Title: _____

a. How many more animals are fish than reptiles? _____

b. How many more fish and mammals are there than birds and reptiles?

_____

HORIZONTAL AND VERTICAL BAR GRAPHS (CONCEPT EXPLORATION TEMPLATE)

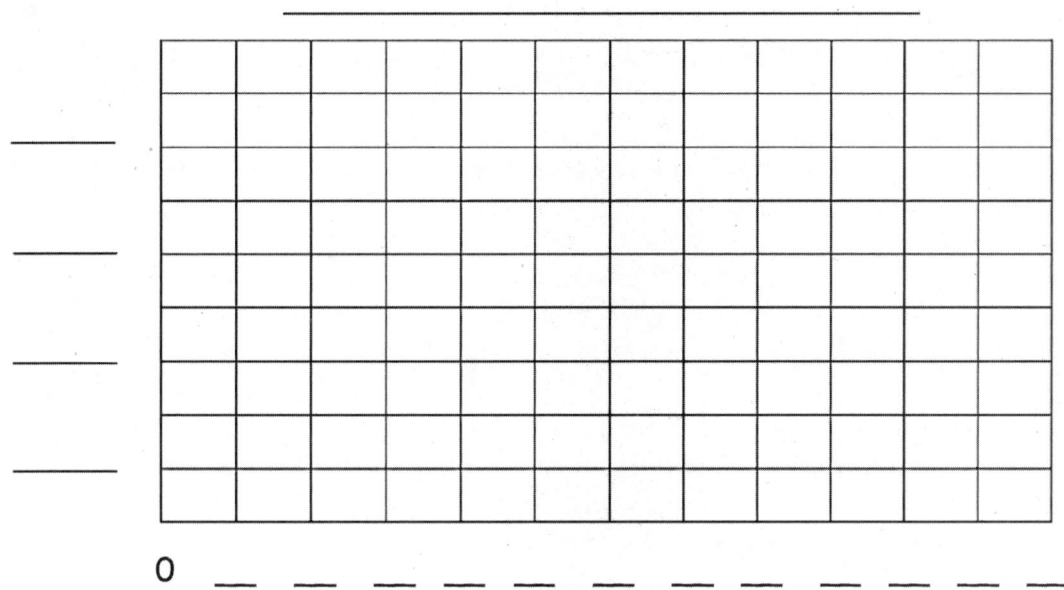

Title: _____

Lesson 4

Word Problem

After a trip to the zoo, Ms. Anderson's students voted on their favorite animals. Use the bar graph to answer the following questions.

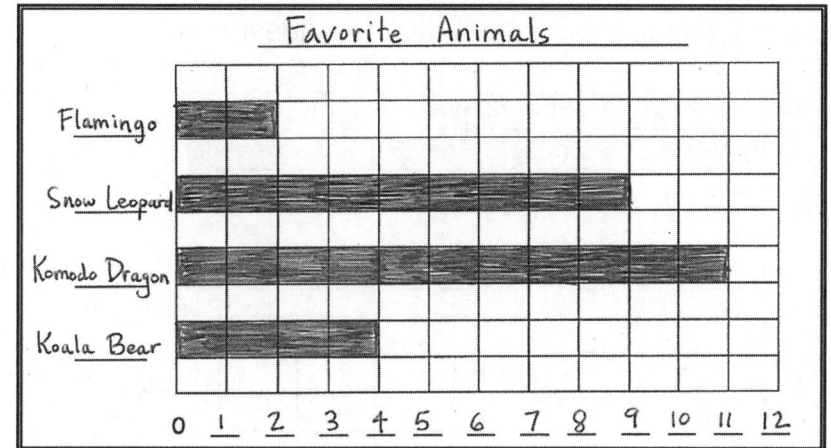

a. Which animal got the fewest votes? _____

b. Which animal got the most votes? _____

c. How many more students liked Komodo dragons than koala bears?

d. Later, two students changed their votes from koala bear to snow leopard. What was the difference between koala bears and snow leopards then?

Name: _____ **Date:** _____

GRADE 2 / MISSION 7 / LESSON 4
Exit Ticket

1. Complete the bar graph using the table with the types of bugs Jeremy counted in his backyard. Then, answer the following questions.

Types of Bugs			
Butterflies	Spiders	Bees	Grasshoppers
4	8	10	9

Title: _____

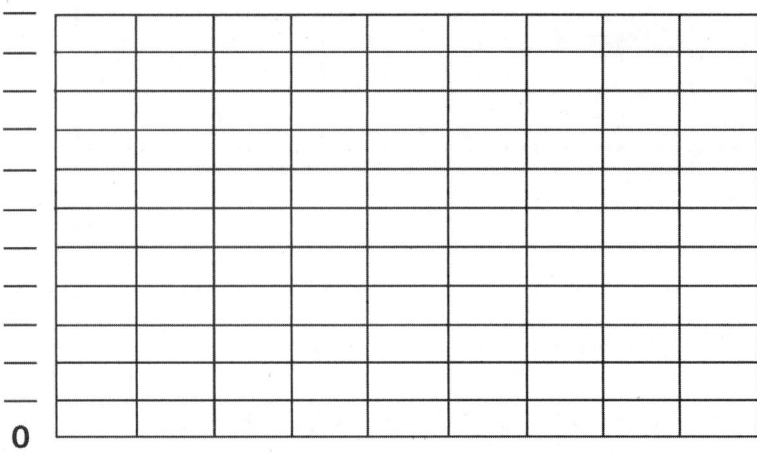

   a. How many more spiders and grasshoppers were counted than bees and butterflies?

   _____

   b. If 5 more butterflies were counted, how many bugs would have been counted?

   _____

HORIZONTAL AND VERTICAL BAR GRAPHS (CONCEPT EXPLORATION TEMPLATE)

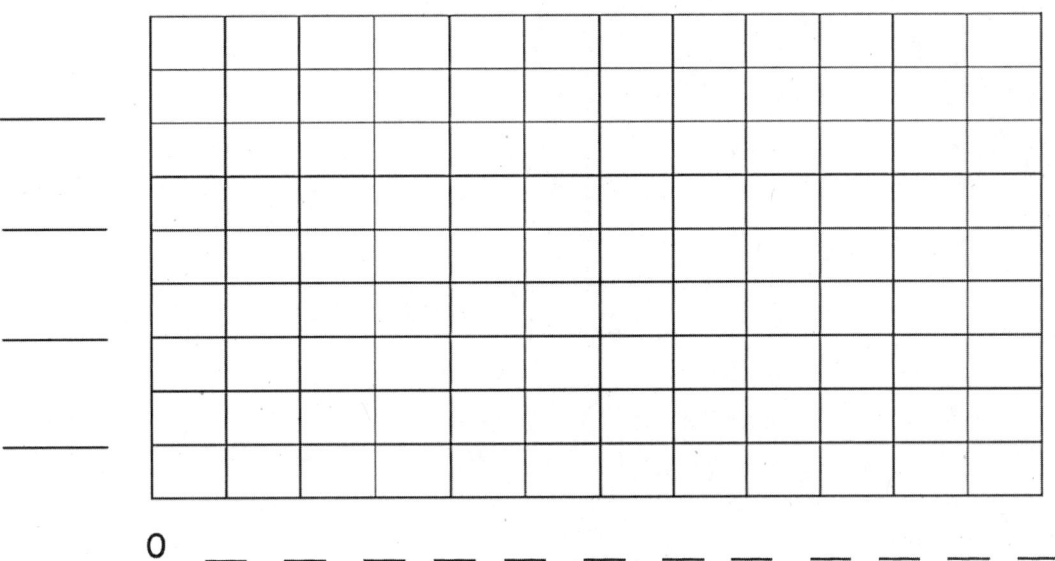

Title: _____

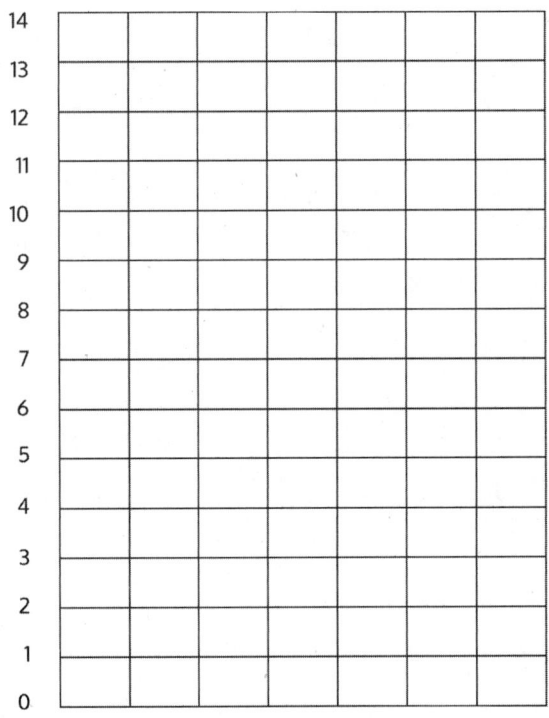

Lesson 5

Word Problem

Rita has 19 more pennies than Carlos. Rita has 27 pennies. How many pennies does Carlos have?

ZEARN MATH Student Edition G2M7 | Lesson 5

Name: _____   Date: _____

GRADE 2 / MISSION 7 / LESSON 5
Exit Ticket

1. Use the table to complete the bar graph. Then, answer the following questions.

Number of Dimes			
Lacy	Sam	Stefanie	Amber
6	11	9	14

Title: _____

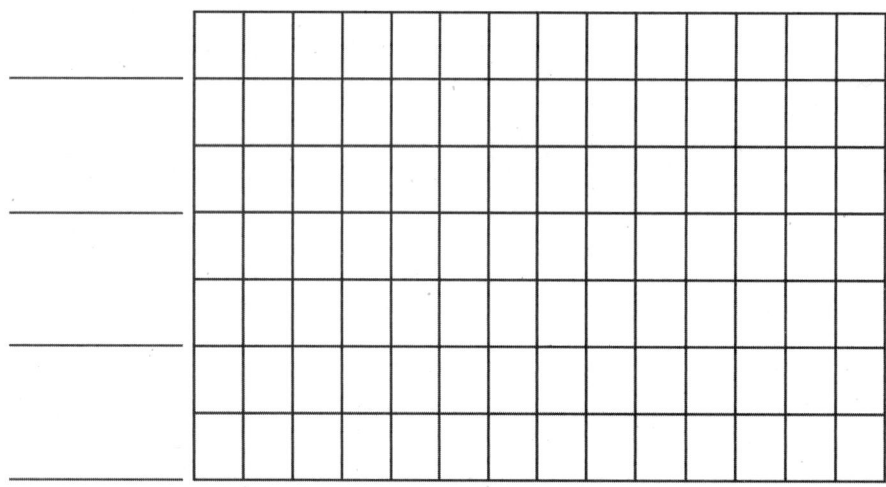

a. How many more dimes does Amber have than Stefanie?

_____

b. How many dimes will Sam and Lacy need to save to equal Stefanie and Amber?

_____

ACTIVITY SHEET 1

Callista saved pennies. Use the table to complete the bar graph. Then, answer the following questions.

Pennies Saved			
Saturday	Sunday	Monday	Tuesday
15	10	4	7

Title: _____

a. How many pennies did Callista save in all? _____

b. Her sister saved 18 fewer pennies. How many pennies did her sister save?

   _____

c. How much more money did Callista save on Saturday than on Monday and Tuesday? _____

d. How will the data change if Callista doubles the amount of money she saved on Sunday? _____

e. Write a comparison question that can be answered using the data on the bar graph.

   _____

ACTIVITY SHEET 2

A group of friends counted their nickels. Use the table to complete the bar graph. Then, answer the following questions.

Amount of Nickels			
Annie	Scarlett	Remy	LaShay
5	11	8	14

Title: _____

0 _ _ _ _ _ _ _ _ _ _ _ _ _ _

a. How many nickels do the children have in all? _____

b. What is the total value of Annie's and Remy's coins? _____

c. How many fewer nickels does Remy have than LaShay?

_____

d. Who has less money, Annie and Scarlett or Remy and LaShay?

_____

e. Write a comparison question that can be answered using the data on the bar graph.

_____

ACTIVITY SHEET 3

1. Design a survey, and collect the data.

2. Label and fill in the table.

3. Use the table to label and complete the bar graph.

4. Write questions based on the graph, and then let students use your graphs to answer them.

   a. _____

   b. _____

   c. _____

   d. _____

Lesson 6

Word Problem

Sarah is saving money in her piggy bank. So far, she has 3 dimes, 1 quarter, and 8 pennies.

a. How much money does Sarah have?

b. How much more does she need to have a dollar?

Name: _____ Date: _____

GRADE 2 / MISSION 7 / LESSON 6
Exit Ticket

Count or add to find the total value of each group of coins.

Write the value using the ¢ or $ symbol.

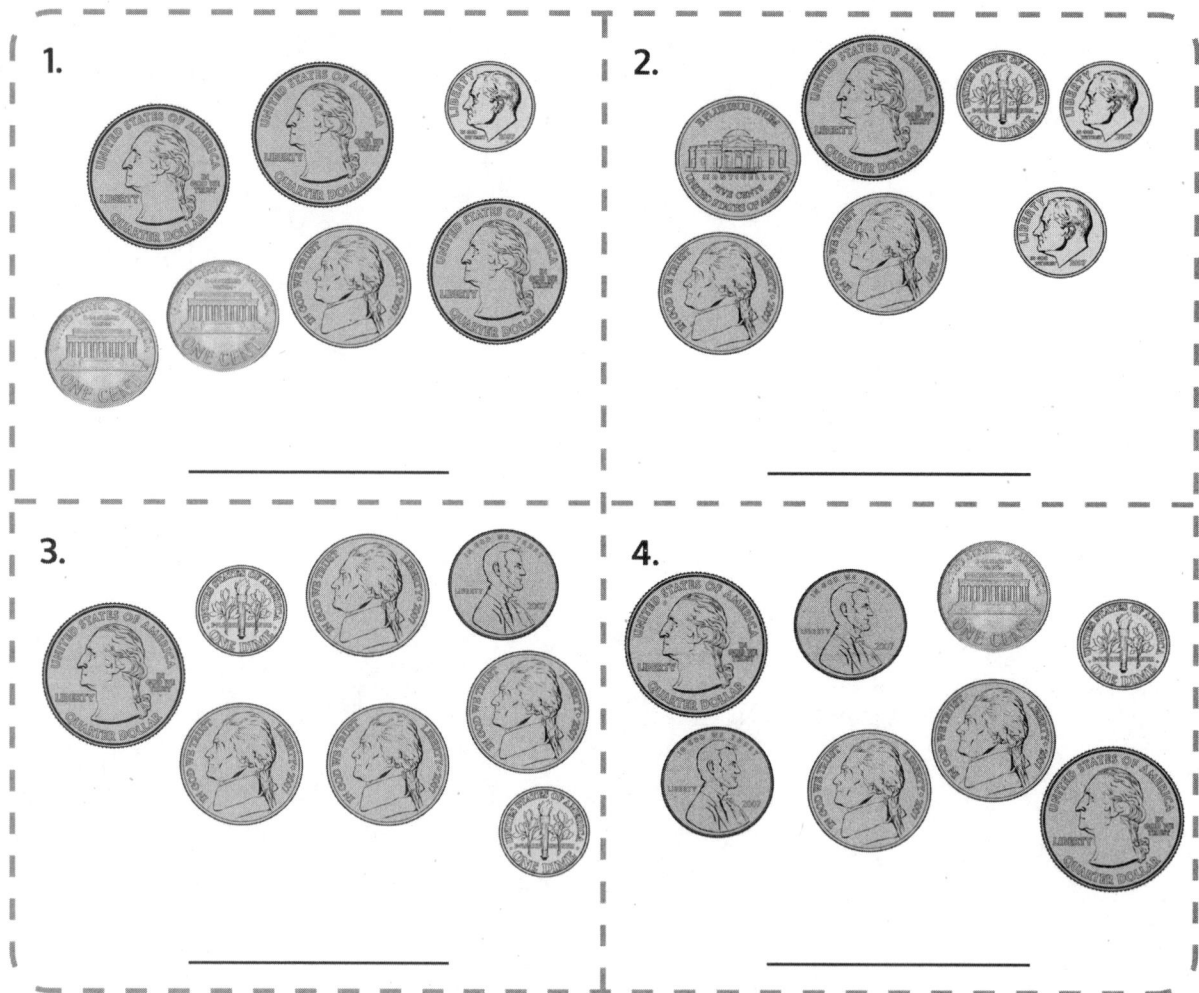

DECOMPOSITION TREE (FLUENCY TEMPLATE)

Lesson 7

Word Problem

Danny has 2 dimes, 1 quarter, 3 nickels, and 5 pennies.

a. What is the total value of Danny's coins?

b. Show two different ways that Danny might add to find the total.

Name: _____   Date: _____

GRADE 2 / MISSION 7 / LESSON 7
Exit Ticket

1. Greg had 1 quarter, 1 dime, and 3 nickels in his pocket. He found 3 nickels on the sidewalk.

   How much money does Greg have?

2. Robert gave Sandra 1 quarter, 5 nickels, and 2 pennies. Sandra already had 3 pennies and 2 dimes.

   How much money does Sandra have now?

Lesson 8

Word Problem

Kiko's brother says that he will trade her 2 quarters, 4 dimes, and 2 nickels for a one-dollar bill. Is this a fair trade? How do you know?

Name: _____ **Date:** _____

GRADE 2 / MISSION 7 / LESSON 8
Exit Ticket

Solve.

1. Josh had 3 five-dollar bills, 2 ten-dollar bills, and 7 one-dollar bills. He gave Suzy 1 five-dollar bill and 2 one-dollar bills. How much money does Josh have left?

2. Jeremy has 3 one-dollar bills and 1 five-dollar bill. Jessica has 2 ten-dollar bills and 2 five-dollar bills. Sam has 2 ten-dollar bills and 4 five-dollar bills. How much money do they have together?

Lesson 9

Word Problem

Clark has 3 ten-dollar bills and 6 five-dollar bills. He has 2 more ten-dollar bills and 2 more five-dollar bills than Shannon. How much money does Shannon have?

Name: _____ **Date:** _____

GRADE 2 / MISSION 7 / LESSON 9
Exit Ticket

1. Smith has 88 pennies in his piggy bank. Write two other coin combinations he could have that would equal the same amount.

Lesson 10

Word Problem

Andrew, Brett, and Jay each have 1 dollar in change in their pockets. They each have a different combination of coins. What coins might each student have in their pocket?

Name: _____   **Date:** _____

GRADE 2 / MISSION 7 / LESSON 10
Exit Ticket

1. Show 36 cents two ways. Use the fewest possible coins on the right below.

	Fewest coins:

2. Show 74 cents two ways. Use the fewest possible coins on the right below.

	Fewest coins:

Lesson 11

Word Problem

Tracy has 85 cents in her change purse. She has 4 coins.

a. Which coins are they?

b. How much more money will Tracy need if she wants to buy a bouncy ball for $1?

Name: _____   Date: _____

GRADE 2 / MISSION 7 / LESSON 11
Exit Ticket

1. 100¢ − 46¢ = _____

2. _____ + 64¢ = 100¢

3. _____ + 13 cents = 100 cents

Lesson 12

Word Problem

Richie has 24 cents. How much more money does he need to make $1?

Name: _____   Date: _____

GRADE 2 / MISSION 7 / LESSON 12
Exit Ticket

Solve using the arrow way, a number bond, or a tape diagram.

1. Jacob bought a piece of gum for 26 cents and a newspaper for 61 cents. He gave the cashier $1. How much money did he get back?

Lesson 13

Word Problem

Dante had some money in a jar. He puts 8 nickels into the jar. Now he has 100 cents. How much money was in the jar at first?

Name: _____   Date: _____

GRADE 2 / MISSION 7 / LESSON 13
Exit Ticket

Solve with a tape diagram and number sentence.

1. Gary went to the store with 4 ten-dollar bills, 3 five-dollar bills, and 7 one-dollar bills. He bought a sweater for $26. What bills did he leave the store with?

Lesson 14

Word Problem

Frances is moving the furniture in her bedroom. She wants to move the bookcase to the space between her bed and the wall, but she is not sure it will fit.

Talk with a partner: What could Frances use as a measurement tool if she doesn't have a ruler? How could she use it?

Show your thinking using pictures, numbers, or words.

Name: _____ **Date:** _____

GRADE 2 / MISSION 7 / LESSON 14
Exit Ticket

Measure the lines below with an inch tile.

Line A _____

   Line A is about _____ inches.

Line B _____

   Line B is about _____ inches.

Line C _____

   Line C is about _____ inches.

PROBLEM SET

1. Measure the objects below with an inch tile. Record the measurements in the table provided.

Object	Measurement
Pair of scissors	
Marker	
Pencil	
Eraser	
Length of worksheet	
Width of worksheet	
Length of desk	
Width of desk	

2. Mark and Melissa both measured the same marker with an inch tile but came up with different lengths. Circle the student work that is correct, and explain why you chose that work.

Melissa's Work

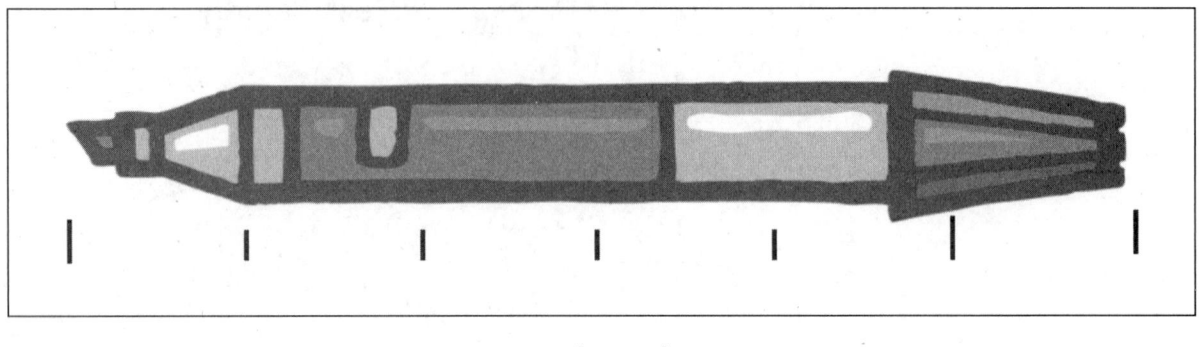

__6__ in

Mark's Work

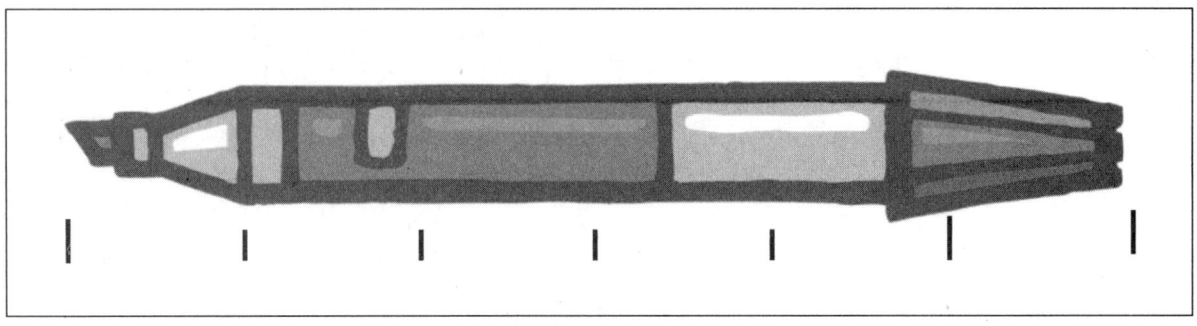

__7__ in

Explanation:

_____

_____

_____

Lesson 15

Word Problem

Edwin and Tina have the same toy truck. Edwin says his is 4 toothpicks long. Tina says hers is 12 lima beans long. How can they both be right?

Work with a partner to measure your object. Partner A, measure with lima beans. Partner B, measure with toothpicks. Use words or pictures to explain how Edwin and Tina can both be right.

Name: _____   Date: _____

GRADE 2 / MISSION 7 / LESSON 15
Exit Ticket

1. Measure and label the sides of the shape below.

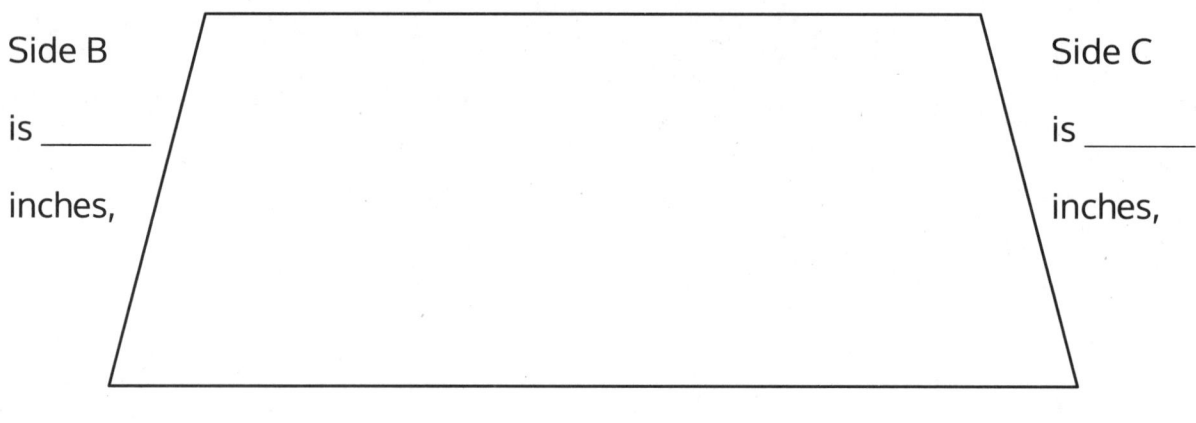

Side A is _____ inches,

Side B is _____ inches,

Side C is _____ inches,

Side D is _____ inches,

2. What is the sum of the length of Side B and the length of Side C?

   _____ inches

Lesson 16

Name: _____   Date: _____

GRADE 2 / MISSION 7 / LESSON 16
Exit Ticket

1. Circle the unit that would best measure each object.

Marker	inch / foot / yard
Height of a car	inch / foot / yard
Birthday card	inch / foot / yard
Soccer field	inch / foot / yard
Length of a computer screen	inch / foot / yard
Height of a bunk bed	inch / foot / yard

RECORDING SHEET

Center 1: Measure and Compare Pencil Lengths

Choose a measuring unit to measure the pencil of everyone in your group. Measure from the bottom of the pencil or eraser to the point of the pencil.

I chose to measure using _____.

Record the results in the table below. Include the units.

Name	Length of Pencil

What is the difference in length between the longest and shortest pencils? Write a number sentence and statement to show the difference between the two lengths.

Center 2: Compare Lengths to a Yardstick

Fill in your estimate for each object using the words *more than*, *less than*, or *about the same length as*. Then, measure each object with a yardstick, and record the measurement on the chart.

Object	Measurement
Length of book	
Height of door	
Length of student desk	

1. The length of a book is _____ the yardstick.

2. The height of the door is _____ the yardstick.

3. The length of a student desk is _____ the yardstick.

What is the length of 4 student desks pushed together with no gaps in between? Use the RDW process to solve.

Center 3: Choose the Units to Measure Objects

Name 4 objects in the classroom. Circle which unit you would use to measure each item, and record the measurement in the chart.

Object	Length of the Object
	inches/feet/yards
	inches/feet/yards
	inches/feet/yards
	inches/feet/yards

Billy measures his pencil. He tells his teacher it is 7 feet long. Explain how you know that Billy is incorrect and how he can change his answer to be correct.

Center 4: Find Benchmarks

Look around the room to find 2 or 3 objects for each benchmark length. Write each object in the chart, and record the exact length.

Objects That Are About an **Inch**	Objects That Are About a **Foot**	Objects That Are About a **Yard**
1. _____ inches	1. _____ inches	1. _____ inches
2. _____ inches	2. _____ inches	2. _____ inches
3. _____ inches	3. _____ inches	3. _____ inches

Center 5: Choose a Tool to Measure

Circle the tool used to measure each object. Then, measure and record the length in the chart. Circle the unit.

Object	Measurement Tool	Measurement
Length of the rug	12-inch ruler / yardstick	_____ inches/feet
Textbook	12-inch ruler / yardstick	_____ inches/feet
Pencil	12-inch ruler / yardstick	_____ inches/feet
Length of the chalkboard	12-inch ruler / yardstick	_____ inches/feet
Pink eraser	12-inch ruler / yardstick	_____ inches/feet

Sera's jump rope is the length of 6 textbooks. Draw a tape diagram to show the length of Sera's jump rope. Then, write a repeated addition sentence using the textbook measurement from the chart to find the length of Sera's jump rope.

Lesson 17

Word Problem

Benjamin measures his forearm and records the length as 15 inches. Then, he measures his upper arm and realizes it's the same!

a. How long is one of Benjamin's arms?

b. What is the total length of both of Benjamin's arms together?

Name: _____   **Date:** _____

GRADE 2 / MISSION 7 / LESSON 17
Exit Ticket

1. Estimate the length of each item by using a mental benchmark. Then, measure the item using feet, inches, or yards.

Item	Mental Benchmark	Estimation	Actual Length
a. Length of an eraser			
b. Width of this paper			

Lesson 18

Word Problem

Ezra is measuring things in his bedroom. He thinks his bed is about 2 yards long. Is this a reasonable estimate? Explain your answer using pictures, words, or numbers.

Name: _____   **Date:** _____

GRADE 2 / MISSION 7 / LESSON 18
Exit Ticket

Measure the lines in inches and centimeters. Round the measurements to the nearest inch or centimeter.

1. _____

   _____ cm _____ in

2. _____

   _____ cm _____ in

Lesson 19

Word Problem

Katia is hanging decorative lights. The strand of lights is 46 feet long. The building wall is 84 feet long. How many more feet of lights does Katia need to buy to equal the length of the wall?

ZEARN MATH Student Edition G2M7 | Lesson 19

Name: _____ Date: _____

GRADE 2 / MISSION 7 / LESSON 19
Exit Ticket

Measure the set of lines in inches and write the length on the line. Complete the comparison sentence.

Line A _____

Line B _____

1. Line A measured about _____ inches.

2. Line B measured about _____ inches.

3. Line A is about _____ inch(es) **longer / shorter** than Line B.

Lesson 20

Name: _____   Date: _____

GRADE 2 / MISSION 7 / LESSON 20
Exit Ticket

Solve using a tape diagram. Use a symbol for the unknown.

1. Jasmine has a jump rope that is 84 inches long. Marie's is 13 inches shorter than Jasmine's. What is the length of Marie's jump rope?

PROBLEM SET

Solve using tape diagrams. Use a symbol for the unknown.

1. Mr. Ramos has knitted 19 inches of a scarf he wants to be 1 yard long. How many more inches of scarf does he need to knit?

2. In the 100-yard race, Jackie has run 76 yards. How many more yards does she have to run?

3. Frankie has a 64-inch piece of rope and another piece that is 18 inches shorter than the first. What is the total length of both ropes?

4. Maria had 96 inches of ribbon. She used 36 inches to wrap a small gift and 48 inches to wrap a larger gift. How much ribbon did she have left?

5. The total length of all three sides of a triangle is 96 feet. The triangle has two sides that are the same length. One of the equal sides measures 40 feet. What is the length of the side that is not equal?

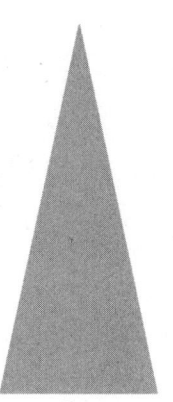

?

6. The length of one side of a square is 4 yards. What is the combined length of all four sides of the square?

Lesson 21

Word Problem

To ride the Mega Mountain roller coaster, riders must be at least 44 inches tall. Caroline is 57 inches tall. She is 18 inches taller than Addison.

How tall is Addison?

How many more inches must Addison grow to ride the roller coaster?

Name: _____   Date: _____

GRADE 2 / MISSION 7 / LESSON 21
Exit Ticket

Find the value of the point on each number line marked by a letter.

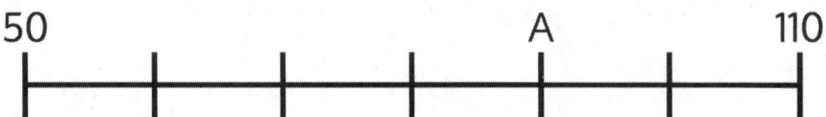

1. Each unit has a length of _____ centimeters.

   A = _____

2. What is the difference between the two endpoints? _____

   B = _____

Lesson 22

Word Problem

Liza, Cecilia, and Dylan are playing soccer. Liza and Cecilia are 120 feet apart. Dylan is in between them. If Dylan is standing the same distance from Liza and Cecilia, how many feet is Dylan from Liza?

ZEARN MATH Student Edition   G2M7 | Lesson 22

Name: _____   Date: _____

GRADE 2 / MISSION 7 / LESSON 22
Exit Ticket

Each unit length on both number lines is 20 centimeters.

(NOTE: Number lines are not drawn to scale.)

1. Show 20 centimeters more than 25 centimeters on the number line.

2. Show 40 centimeters less than 45 centimeters on the number line.

3. Write an addition or a subtraction sentence to match each number line.

NUMBER LINES A AND B (CONCEPT EXPLORATION TEMPLATE)

Number Line A

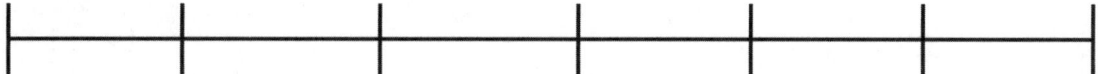

Number Line B

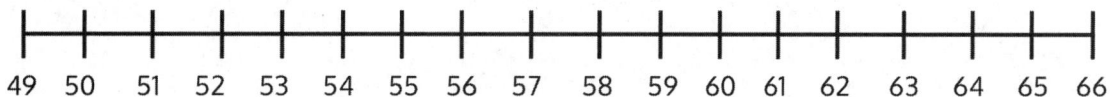

Lesson 23

Name: _____   Date: _____

GRADE 2 / MISSION 7 / LESSON 23
Exit Ticket

1. The lines below have been measured for you. Record the data using tally marks on the table provided and answer the following question.

   Line A   5 inches
   _____

   Line B   6 inches
   _____

   Line C   4 inches
   _____

   Line D   6 inches
   _____

   Line E   3 inches
   _____

Line Length	Number of Lines
Shorter than 5 inches	
5 inches or longer	

2. If 8 more lines were measured to be longer than 5 inches and 12 more lines were measured to be shorter than 5 inches, how many tallies would be in the chart?

   _____ tallies would be in the chart.

RECORDING SHEET (CONCEPT EXPLORATION TEMPLATE)

1. Gather and record group data.

   Write your teacher's handspan measurement here: _____

   Measure your handspan, and record the length here: _____

   Measure the handspans of the other people in your group, and write them here. We will be using the data tomorrow.

   Name: **Handspan:**

   _____   _____

   _____   _____

   _____   _____

   _____   _____

Handspan	Tally of Number of People
3 inches	
4 inches	
5 inches	
6 inches	
7 inches	
8 inches	

   What is the most common handspan length? _____

   What is the least common handspan length? _____

   What do you think the most common handspan length will be for the whole class? Explain why.

   _____

   _____

   _____

   _____

2. Record the entire class data.

   Record the data using tally marks on the table provided.

Handspan	Tally of Number of People
3 inches	
4 inches	
5 inches	
6 inches	
7 inches	
8 inches	

   What handspan length is the most common? _____

   What handspan length is the least common? _____

   Ask and answer a comparison question that can be answered using the data above.

   Question: _____

   _____

   Answer: _____

   _____

Lesson 24

Word Problem

Mike, Dennis, and April all collected coins from a parking lot. When they counted their coins, they had 24 pennies, 15 nickels, 7 dimes, and 2 quarters. They put all the pennies into one cup and the other coins in another. Which cup has more coins? How many more?

GRADE 2 / MISSION 7 / LESSON 24
Exit Ticket

1. Use the data in the table to create a line plot.

   Length of Crayons in a Class Bin

Crayon Length (inches)	Number of crayons
1				
2	⦀⦀ ||||
3	⦀⦀ ||
4	⦀⦀

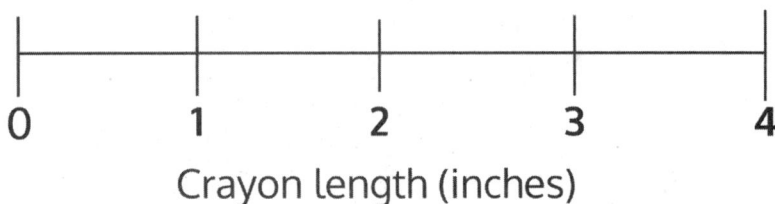

Crayon length (inches)

RECORDING SHEET (CONCEPT EXPLORATION TEMPLATE)

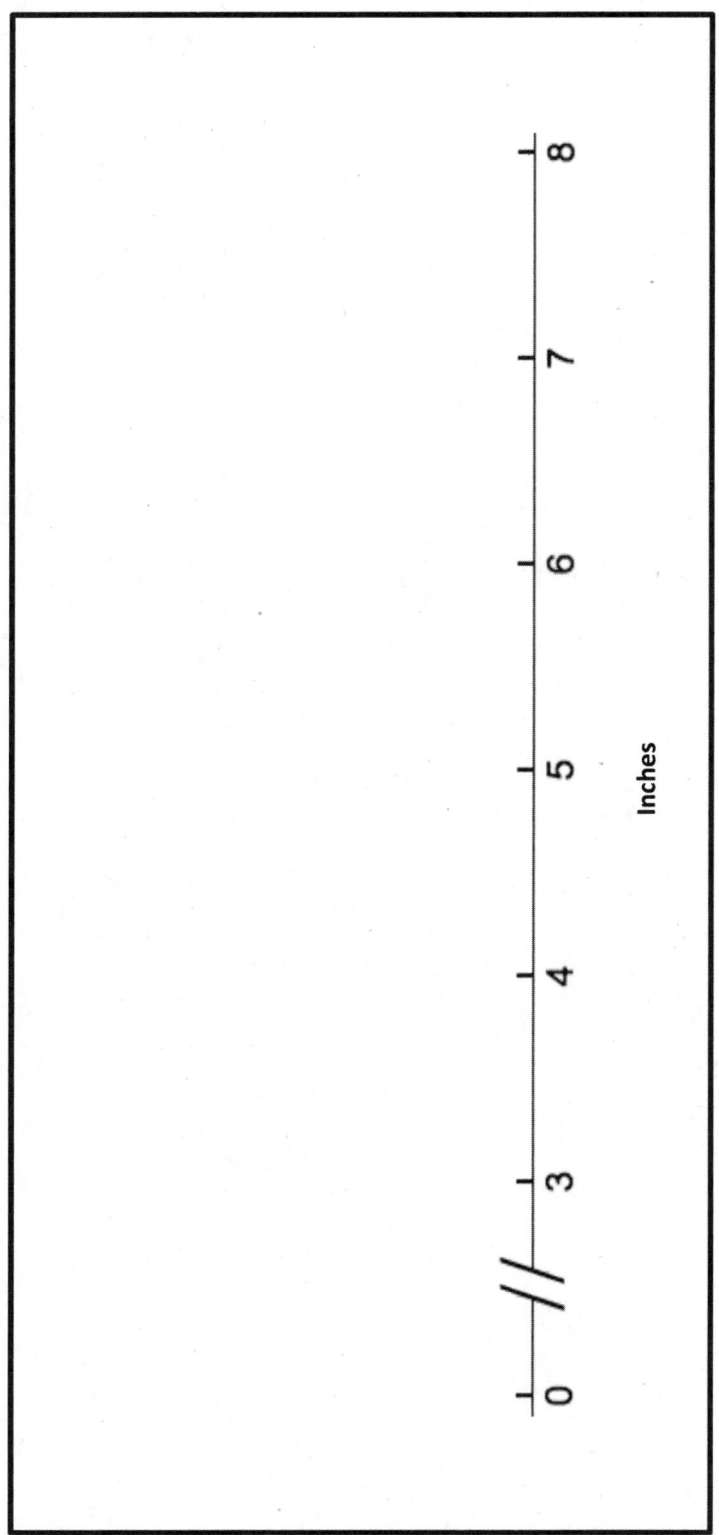

Lesson 25

Word Problem

These are the types and numbers of stamps in Shannon's stamp collection.

Type of Stamp	Number of Stamps
Holiday	16
Animal	8
Birthday	9
Famous singers	21

Her friend Michael gives her some flag stamps. If he gives her 7 fewer flag stamps than birthday and animal stamps together, how many flag stamps does she have?

Extension: If the flag stamps are worth 12 cents each, what is the total value of Shannon's flag stamps?

Name: _____ **Date:** _____

GRADE 2 / MISSION 7 / LESSON 25
Exit Ticket

Answer the questions using the line plot below.

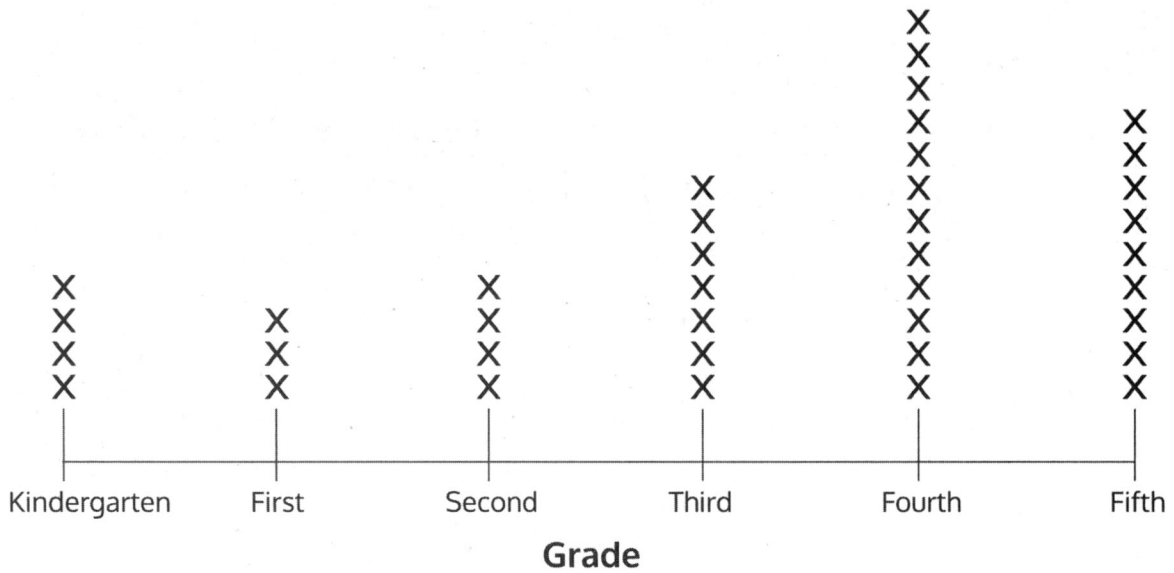

1. How many students went to the baseball game? _____

2. What is the difference between the number of first-grade students and the number of fourth-grade students who went to the baseball game?

   _____

3. Come up with a possible explanation for why most of the students who attended are in the upper grades.

   _____
   _____

Lesson 26

Word Problem

Judy bought a music player and a set of earphones. The earphones cost $9, which is $48 less than the music player. How much change should Judy get back if she gave the cashier a $100 bill?

Name: _____   Date: _____

GRADE 2 / MISSION 7 / LESSON 26
Exit Ticket

Use the data in the table provided to create a line plot.

The table below describes the heights of second-grade students on the soccer team.

Height (inches)	Number of Students
35	3
36	4
37	7
38	8
39	6
40	5

LENGTH AND TEMPERATURE TABLES (CONCEPT EXPLORATION TEMPLATE 1)

Length of Items in Our Pencil Boxes	Number of Items
6 cm	1
7 cm	2
8 cm	4
9 cm	3
10 cm	6
11 cm	4
13 cm	1
16 cm	3
17 cm	2

Temperatures in May	Number of Days
59°	1
60°	3
63°	3
64°	4
65°	7
67°	5
68°	4
69°	3
72°	1

GRID PAPER (CONCEPT EXPLORATION TEMPLATE 2)

GRID PAPER (CONCEPT EXPLORATION TEMPLATE 2)

THERMOMETER (CONCEPT EXPLORATION TEMPLATE 3)

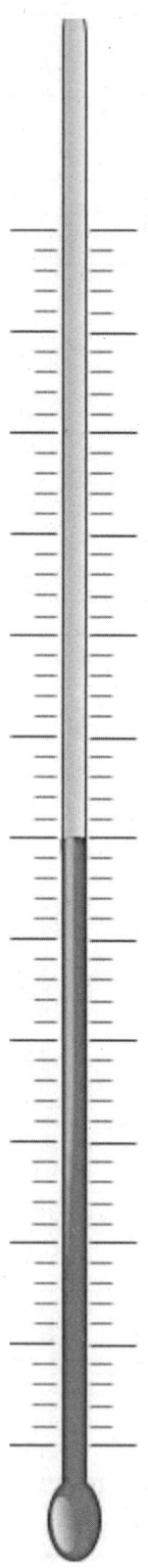

Grade 2

Mission 8

Shapes, Time, and Fractions

Lesson 1

Word Problem

Terrence is making shapes with 12 toothpicks. Using all of the toothpicks, create 3 different shapes he could make. How many other combinations can you find?

GRADE 2 / MISSION 8 / LESSON 1
Exit Ticket

Study the shapes below. Then answer the questions.

A 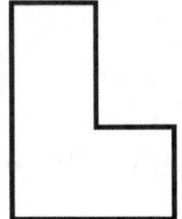 B

C 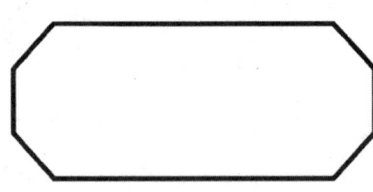 D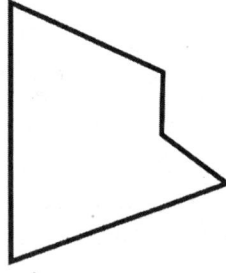

1. Which shape has the most sides? _____

2. Which shape has 3 fewer angles than shape C? _____

3. What shape has 3 more sides than shape B? _____

4. Which of these shapes have the same number of sides as angles? _____

Lesson 2

Word Problem

How many triangles can you find? (Hint: If you only found 10, keep looking!)

Name: _____ **Date:** _____

GRADE 2 / MISSION 8 / LESSON 2
Exit Ticket

1. Count the number of sides and angles for each shape to identify each polygon. The polygon names in the word bank may be used more than once.

   | Hexagon   Quadrilateral   Triangle   Pentagon |

   a. _____

   b. _____

   c. _____

   d. _____

   e. _____

   f. _____

Lesson 3

Word Problem

Three sides of a quadrilateral have the following lengths: 19 cm, 23 cm, and 26 cm. If the total distance around the shape is 86 cm, what is the length of the fourth side?

Name: _____ Date: _____

GRADE 2 / MISSION 8 / LESSON 3
Exit Ticket

1. Use a straight edge to draw the polygon with the given attributes in the space below.

    Draw a five-sided polygon.

    Number of angles: _____

    Name of polygon: _____

HUNDREDS PLACE VALUE CHART (FLUENCY TEMPLATE)

Ones
Tens
Hundreds

Workspace:

CORE FLUENCY PRACTICE SET A

1.	10 + 9 =	21.	3 + 9 =
2.	10 + 1 =	22.	4 + 8 =
3.	11 + 2 =	23.	5 + 9 =
4.	13 + 6 =	24.	8 + 8 =
5.	15 + 5 =	25.	7 + 5 =
6.	14 + 3 =	26.	5 + 8 =
7.	13 + 5 =	27.	8 + 3 =
8.	12 + 4 =	28.	6 + 8 =
9.	16 + 2 =	29.	4 + 6 =
10.	18 + 1 =	30.	7 + 6 =
11.	11 + 7 =	31.	7 + 4 =
12.	13 + 4 =	32.	7 + 9 =
13.	14 + 5 =	33.	7 + 7 =
14.	9 + 4 =	34.	8 + 6 =
15.	9 + 2 =	35.	6 + 9 =
16.	9 + 9 =	36.	8 + 5 =
17.	6 + 9 =	37.	4 + 7 =
18.	8 + 9 =	38.	3 + 9 =
19	7 + 8 =	39.	8 + 6 =
20.	8 + 8 =	40.	9 + 4 =

CORE FLUENCY PRACTICE SET B

1.	10 + 8 =		21.	5 + 8 =
2.	4 + 10 =		22.	6 + 7 =
3.	9 + 10 =		23.	___ + 4 = 12
4.	11 + 5 =		24.	___ + 7 = 13
5.	13 + 3 =		25.	6 + ___ = 14
6.	12 + 4 =		26.	7 + ___ = 15
7.	16 + 3 =		27.	___ = 9 + 8
8.	15 + ___ = 19		28.	___ = 7 + 5
9.	18 + ___ = 20		29.	___ = 4 + 8
10.	13 + 5 =		30.	3 + 9 =
11.	___ = 4 + 16		31.	6 + 7 =
12.	___ = 6 + 12		32.	8 + ___ = 13
13.	___ = 14 + 6		33.	___ = 7 + 9
14.	9 + 3 =		34.	6 + 6 =
15.	7 + 9 =		35.	___ = 7 + 5
16.	___ + 4 = 11		36.	___ = 4 + 8
17.	___ + 6 = 13		37.	20 = 13 + ___
18.	___ + 5 = 12		38.	18 = ___ + 9
19.	___ + 8 = 14		39.	16 = ___ + 7
20.	___ + 9 = 15		40.	20 = 9 + ___

CORE FLUENCY PRACTICE SET C

1.	19 − 9 =	21.	15 − 7 =
2.	19 − 11 =	22.	18 − 9 =
3.	17 − 10 =	23.	16 − 8 =
4.	12 − 2 =	24.	15 − 6 =
5.	15 − 12 =	25.	17 − 8 =
6.	18 − 10 =	26.	14 − 6 =
7.	17 − 5 =	27.	16 − 9 =
8.	20 − 9 =	28.	13 − 8 =
9.	14 − 4 =	29.	12 − 5 =
10.	16 − 13 =	30.	19 − 8 =
11.	11 − 2 =	31.	17 − 9 =
12.	12 − 3 =	32.	16 − 7 =
13.	14 − 2 =	33.	14 − 8 =
14.	13 − 4 =	34.	15 − 9 =
15.	11 − 3 =	35.	13 − 7 =
16.	12 − 4 =	36.	12 − 8 =
17.	13 − 2 =	37.	15 − 8 =
18.	14 − 5 =	38.	14 − 9 =
19	11 − 4 =	39.	12 − 7 =
20.	12 − 5 =	40.	11 − 9 =

CORE FLUENCY PRACTICE SET D

1.	12 − 3 =		21.	13 − 7 =
2.	13 − 5 =		22.	15 − 9 =
3.	11 − 2 =		23.	18 − 7 =
4.	12 − 5 =		24.	14 − 7 =
5.	13 − 4 =		25.	17 − 9 =
6.	13 − 2 =		26.	12 − 9 =
7.	11 − 4 =		27.	13 − 6 =
8.	12 − 6 =		28.	15 − 7 =
9.	11 − 3 =		29.	16 − 8 =
10.	13 − 6 =		30.	12 − 6 =
11.	___ = 11 − 9		31.	___ = 13 − 9
12.	___ = 13 − 8		32.	___ = 17 − 8
13.	___ = 12 − 7		33.	___ = 14 − 9
14.	___ = 11 − 6		34.	___ = 13 − 5
15.	___ = 13 − 9		35.	___ = 15 − 8
16.	___ = 14 − 8		36.	___ = 18 − 9
17.	___ = 11 − 7		37.	___ = 16 − 7
18.	___ = 15 − 6		38.	___ = 20 − 12
19.	___ = 16 − 9		39.	___ = 20 − 6
20.	___ = 12 − 8		40.	___ = 20 − 17

CORE FLUENCY PRACTICE SET E

1.	13 − 4 =	21.	8 + 4 =
2.	15 − 8 =	22.	6 + 7 =
3.	19 − 5 =	23.	9 + 9 =
4.	11 − 7 =	24.	12 − 6 =
5.	9 + 6 =	25.	16 − 7 =
6.	7 + 8 =	26.	13 − 5 =
7.	4 + 7 =	27.	11 − 8 =
8.	13 + 6 =	28.	7 + 9 =
9.	12 − 8 =	29.	5 + 7 =
10.	17 − 9 =	30.	8 + 7 =
11.	14 − 6 =	31.	9 + 8 =
12.	16 − 7 =	32.	11 + 9 =
13.	6 + 8 =	33.	12 − 3 =
14.	7 + 6 =	34.	14 − 5 =
15.	4 + 9 =	35.	20 − 13 =
16.	5 + 7 =	36.	8 − 5 =
17.	9 − 5 =	37.	7 + 4 =
18.	13 − 7 =	38.	13 + 5 =
19.	16 − 9 =	39.	7 + 9 =
20.	14 − 8 =	40.	8 + 11 =

Lesson 4

Name: _____   Date: _____

GRADE 2 / MISSION 8 / LESSON 4
Exit Ticket

Use crayons to trace the parallel sides on each quadrilateral. Use your index card to find each square corner, and box it.

1.

2.

3.

4.

Lesson 5

Word Problem

Owen had 90 straws to create pentagons. He created a set of 5 pentagons when he noticed a number pattern. How many more shapes can he add to the pattern?

Name: _____ **Date:** _____

GRADE 2 / MISSION 8 / LESSON 5
Exit Ticket

1. Draw 3 cubes. Put a star next to your best one.

Lesson 6

Word Problem

Frank has 19 fewer cubes than Josie. Frank has 56 cubes. They want to use all of their cubes to build a tower. How many cubes will they use?

Name: _____   Date: _____

GRADE 2 / MISSION 8 / LESSON 6
Exit Ticket

Use your tangram pieces to make two new polygons. Draw a picture of each new polygon, and name them.

1.

2.

Lesson 7

Word Problem

Mrs. Libarian's students are picking up tangram pieces. They collect 13 parallelograms, 24 large triangles, 24 small triangles, and 13 medium triangles. The rest are squares. If they collect 97 pieces in all, how many squares are there?

Name: _____ Date: _____

GRADE 2 / MISSION 8 / LESSON 7
Exit Ticket

1. Circle the shapes that show thirds.

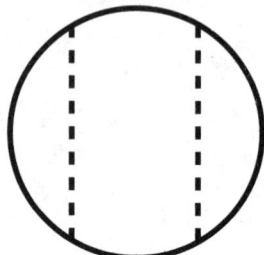 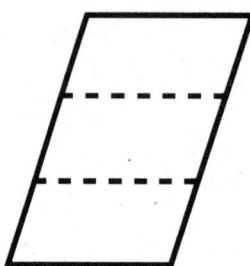 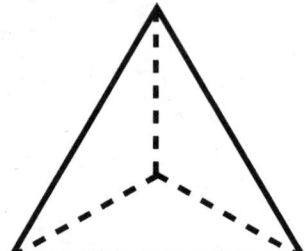

2. Circle the shapes that show fourths.

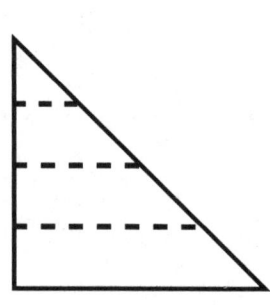 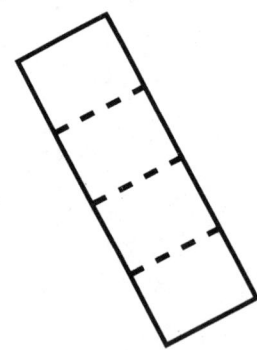 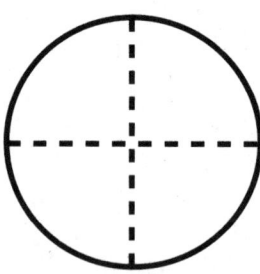

Lesson 8

Word Problem

Students were making larger shapes out of triangles and squares. They put away all 72 triangles. There were still 48 squares on the carpet. How many triangles and squares were on the carpet when they started?

Name: _____ Date: _____

GRADE 2 / MISSION 8 / LESSON 8
Exit Ticket

1. Name a shape that would cover half of this rectangle:

   _____

PROBLEM SET

1. Use one pattern block to cover half the rhombus.

   a. Identify the pattern block used to cover half of the rhombus. _____

   b. Draw a picture of the rhombus formed by the 2 halves.

2. Use one pattern block to cover half the hexagon.

   a. Identify the pattern block used to cover half of a hexagon. _____

   b. Draw a picture of the hexagon formed by the 2 halves.

3. Use one pattern block to cover 1 third of the hexagon.

   a. Identify the pattern block used to cover 1 third of a hexagon. _____

   b. Draw a picture of the hexagon formed by the 3 thirds.

4. Use one pattern block to cover 1 third of the trapezoid.

   a. Identify the pattern block used to cover 1 third of a trapezoid. _____

   b. Draw a picture of the trapezoid formed by the 3 thirds.

5. Use 4 pattern block squares to make one larger square.

   a. Draw a picture of the square formed in the space below.

   b. Shade 1 small square. Each small square is 1 _____ (half / third / fourth) of the whole square.

   c. Shade 1 more small square. Now, 2 _____ (halves / thirds / fourths) of the whole square is shaded.

   d. 2 fourths of the square is the same as 1 _____ (half / third / fourth) of the whole square.

   e. Shade 2 more small squares. _____ fourths is equal to 1 whole.

6. Use one pattern block to cover 1 sixth of the hexagon.

   a. Identify the pattern block used to cover 1 sixth of a hexagon. _____

   b. Draw a picture of the hexagon formed by the 6 sixths.

Lesson 9

Word Problem

Mr. Thompson's class raised 96 dollars for a field trip. They need to raise a total of 120 dollars.

a. How much more money do they need to raise in order to reach their goal?

b. If they raise 86 more dollars, how much extra money will they have?

Name: _____ **Date:** _____

GRADE 2 / MISSION 8 / LESSON 9
Exit Ticket

Shade 1 half of the shapes that are split into 2 equal shares.

a.

b.

c.

d.

e.

f.

g.

Lesson 10

Word Problem

Felix is passing out raffle tickets. He passes out 98 tickets and has 57 left. How many raffle tickets did he have to start?

ZEARN MATH Student Edition G2M8 | Lesson 10

Name: _____ Date: _____

GRADE 2 / MISSION 8 / LESSON 10
Exit Ticket

1. Partition and shade the following shapes as indicated. Each rectangle or circle is one whole.

   a. 2 halves

   b. 2 thirds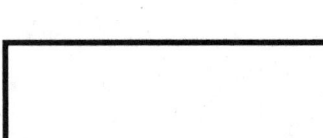

   c. 1 third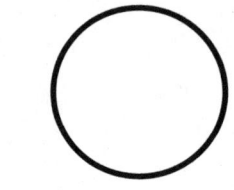

   d. 1 half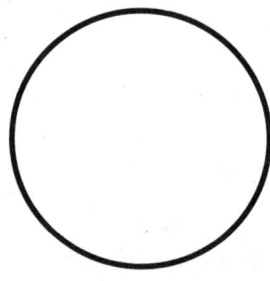

   e. 2 fourths

   f. 1 fourth

Lesson 11

Word Problem

Jacob collected 70 baseball cards. He gave half of them to his brother, Sammy. How many baseball cards does Jacob have left?

ZEARN MATH Student Edition   G2M8 | Lesson 11

Name: _____   Date: _____

GRADE 2 / MISSION 8 / LESSON 11
Exit Ticket

1. What fraction do you need to color so that 1 whole is shaded?

   a.

   _____

   b.

   _____

   c.

   _____

   d.

   _____

248   G2M8 | Lesson 11

Lesson 12

Word Problem

Tugu made two pizzas for himself and his 5 friends to share. He wants everyone to have an equal share of the pizza. Should he cut the pizzas into halves, thirds, or fourths?

ZEARN MATH Student Edition   G2M8 | Lesson 12

Name: _____   Date: _____

GRADE 2 / MISSION 8 / LESSON 12
Exit Ticket

1. Partition the rectangle in 2 different ways to show equal shares.

    a. 2 halves

    b. 3 thirds

    c. 4 fourths

PROBLEM SET

1. Partition the rectangles in 2 different ways to show equal shares.

   a. 2 halves

   ☐   ☐

   b. 3 thirds

   ☐   ☐

   c. 4 fourths

   ☐   ☐

2. Build the original whole square using the rectangle half and the half represented by your 4 small triangles. Draw it in the space below.

3. Use different-colored halves of a whole square.

    a. Cut the square in half to make 2 equal-size rectangles.

    b. Rearrange the halves to create a new rectangle with no gaps or overlaps.

    c. Cut each equal part in half to make 4 equal-size squares.

    d. Rearrange the new equal shares to create different polygons.

    e. Draw one of your new polygons from Part (d) below.

Extension

4. Cut out the circle.

    a. Cut the circle in half.

    b. Rearrange the halves to create a new shape with no gaps or overlaps.

    c. Cut each equal share in half.

    d. Rearrange the equal shares to create a new shape with no gaps or overlaps.

    e. Draw your new shape from Part (d) below.

Lesson 13

Name: _____   Date: _____

GRADE 2 / MISSION 8 / LESSON 13
Exit Ticket

1. Draw the minute hand on the clock to show the correct time.

Half past 7   12:15   A quarter to 3

Lesson 14

Word Problem

Brownies take 45 minutes to bake. Pizza takes half an hour less than brownies to warm up. How long does pizza take to warm up?

GRADE 2 / MISSION 8 / LESSON 14
Exit Ticket

Draw the hour and minute hands on the clocks to match the correct time.

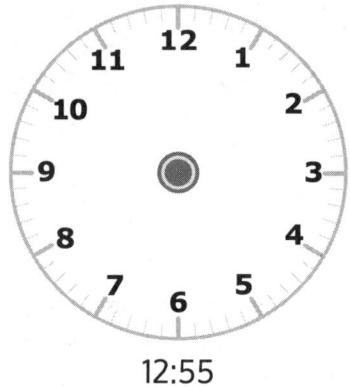

12:55

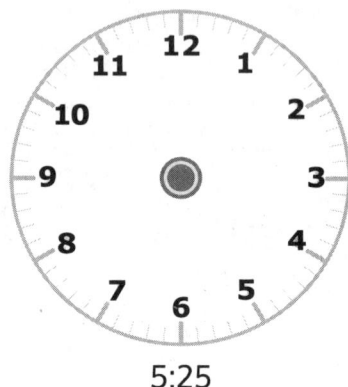

5:25

Lesson 15

Word Problem

At Memorial School, students have a quarter hour for morning recess and 33 minutes for a lunch break.

How much free time do they have in all? How much more time for lunch than recess do they have?

ZEARN MATH Student Edition   G2M8 | Lesson 15

Name: _____   Date: _____

GRADE 2 / MISSION 8 / LESSON 15
Exit Ticket

1. Draw the hands on the analog clock to match the time on the digital clock. Then, circle **a.m.** or **p.m.** based on the description given.

   a. The sun is rising

   6:10 a.m. or p.m.
   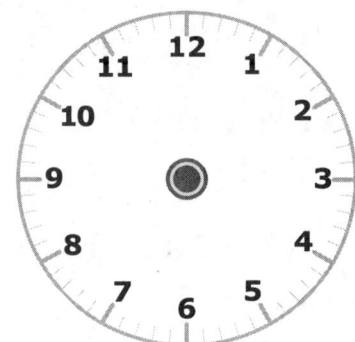

   b. Walking the dog

   3:40 a.m. or p.m.
   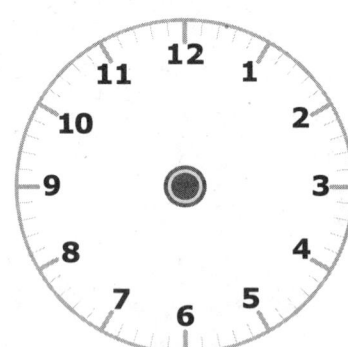

258

TELLING TIME STORY (CONCEPT EXPLORATION TEMPLATE) (PAGE 1 OF 2)

Write the time. Circle a.m. or p.m.

TELLING TIME STORY (CONCEPT EXPLORATION TEMPLATE) (PAGE 2 OF 2)

Write the time. Circle a.m. or p.m.

Lesson 16

Word Problem

On Saturdays, Jean may only watch cartoons for one hour. Her first cartoon lasts 14 minutes, and the second lasts 28 minutes. After a 5-minute break, Jean watches a 15-minute cartoon. How much time does Jean spend watching cartoons? Did she break her time limit?

Name: _____ **Date:** _____

GRADE 2 / MISSION 8 / LESSON 16
Exit Ticket

1. How much time has passed?

   a. 3:00 p.m. → 11:00 p.m.   _____

   b. 5:00 a.m. → 12:00 p.m. (noon)   _____

   c. 9:30 p.m. → 7:30 a.m.   _____